IMAGES
of America

CEMETERIES OF SAN DIEGO COUNTY

THE SAN DIEGO GRAVESTONE PROJECT. San Diego State University students and volunteers record data from grave markers at the All Saints Episcopal Church and Cemetery in Oceanside. The church was established in 1890 by the English colonists of San Luis Rey Valley. Dr. Seth Mallios directs the San Diego Gravestone Project (SDGP), and David M. Caterino, Registered Professional Archaeologist, supervises the fieldwork. The SDGP began in 2002. (Courtesy South Coastal Information Center [SCIC].)

ON THE COVER: Stuart Lake poses with the gravestone of Harry Moore Forster in 1935. Forster died in 1888 and was buried behind the Kimble-Wilson Store in Warner Springs. (Courtesy San Diego Historical Society [SDHS].)

IMAGES
of America

CEMETERIES OF SAN DIEGO COUNTY

David M. Caterino and Seth Mallios

ISBN 978-1-5316-3741-5

Published by Arcadia Publishing
Charleston SC, Chicago IL, Portsmouth NH, San Francisco CA

Library of Congress Catalog Card Number: 2007941839

For all general information contact Arcadia Publishing at:
Telephone 843-853-2070
Fax 843-853-0044
E-mail sales@arcadiapublishing.com
For customer service and orders:
Toll-Free 1-888-313-2665

Visit us on the Internet at www.arcadiapublishing.com

To Spot . . . always an interested observer.

Contents

Acknowledgments

It is the unsung people behind the scenes who make a book like this possible. We would like to thank those individuals who gave us clues, supplied information, allowed access to their land, pointed us in the right direction, prevented wild goose chases, made introductions, prepared maps, or advertised our project. In no particular order and with apologies to those we have mistakenly overlooked, we extend our sincerest appreciation to Scott Mattingly; David Lewis; Kristi Hawthorne; Jacque Beck; Heather Thomson; Carrie Gregory; Jeannie Gregory; Dr. Lynne Christenson; Spike Alford; Michele Clock; John Gibbins; Patrick McGinnis; Debbie Moretti; Nelda Taylor; Robert and Dixie North; Richard Moore; Larry Brooks; Bill Parsons; Joan Miller; Eagle Scout Karl Miller and his troop, family, and friends; Berry-Bell and Hall Mortuary; Larry Brooks; Bill Parsons; James Harper; Ruth Ochoa; Norm and Kathy Feigel; Wayne Taylor; Chet and Trudy Taylor; Edith Swaim; Louis Guassac; Carole Bradford; Al Gettman; Louis Spaulding; James Hinds; Tracy Nelson; Tom and Nancy Wilson; George White; Dan Galligan; Jennifer Miller; Mary Beth Hayashi; Jim Kemp; Wayne Mills; Jai Bastian; Bill Arballo; Hortensia Trejo; Dan McManama; Pat Valenta; Mabel Carlson; Jerry Schaefer; Steve Van Wormer; Andrew Pigniolo; Stan Berryman; Monica Guerrero; Woody Kirkman; Joyce Pelkey; Jan McDonald; and Shane Emmett. In addition, special thanks to the employees of the South Coastal Information Center and the many San Diego State University graduate and undergraduate students who accomplished archival searches, assisted in field investigations, recorded gravestone information, and performed data entry.

INTRODUCTION

This book is a sequel to our 2007 text, *Cemeteries of San Diego*. Whereas the previous book detailed the city's 32 cemeteries, grave sites, and popular mortuary myths, the work presented here focuses on 103 burial sites throughout San Diego County that are located outside of the city limits. Both books have resulted from the San Diego Gravestone Project (SDGP), a comprehensive survey of the region's historical cemeteries and grave markers. Before the inception of the SDGP five years ago, local scholars had recorded 40 county cemeteries; we now know of nearly 100 others.

Unfortunately, many San Diego County burial grounds are no longer clearly marked. In numerous cases, the gravestones that once identified the exact location and identity of the deceased have been moved, destroyed, buried, or otherwise separated from the dead. As a result, local cemeteries are in danger of being lost for eternity. Development, neglect, and vandalism are the obvious culprits. They erode the continuity between San Diegans of the past and those of the present. However, despite repeated graveyard desecration, San Diego County's historical cemeteries have not disappeared entirely from the local landscape. Clues remain as to where early San Diegans are buried. This book serves to gather and record these hints of the past before they vanish as well. As archaeologists, we call these clues "palimpsests." They are tangible aboveground reflections of that which lies below, faint imprints of the past slightly blurred by the sands of time.

There is no single way to find a lost graveyard. In fact, we have employed a variety of sources and techniques. Historical maps and photographs, obscure archaeological reports, ground-penetrating radar surveys, personal accounts, and sometimes a bit of luck have all been useful in our quest to rediscover San Diego's past.

Cemeteries of San Diego County traces the history of the region through its graveyards and discusses them by type. Local cemeteries fall into four general categories: 1) mission cemeteries, 2) California Indian cemeteries, 3) pioneer cemeteries, and 4) large cemeteries. A few of the graveyards fit into more than one category, but most can be included in this classificatory scheme. Although Native Americans occupied the area for 7,000 to 10,000 years before Europeans first visited Southern California in 1542, the prehistoric burials from these indigenous communities are not included in the SDGP and this book for two reasons. First, these pre-contact burials lacked aboveground grave markers. Even though the cremated remains of many early California Indians were often associated with other contemporaneous archaeological features, it does not appear that individual markers exclusively denoted the deceased. Second, following federal and state guidelines, this study endeavors to protect archaeological sites from looting by keeping the exact location of prehistoric burials confidential.

Spanish and Mexican burial practices dominated the early Colonial cemeteries of San Diego. These Catholic graveyards were on consecrated ground in or near the missions. They were often small in size and limited in space, holding the graves of clergy, soldiers, and settlers. Native American converts were buried outside of the cemetery proper. Most of the original graves from San Diego's Spanish and Mexican periods (1769–1821 and 1821–1848 respectively) were marked with simple wooden crosses. Individual grave plots were often surrounded by picket fences as well.

Almost all of the original grave markers from this era are gone, but some have been replaced by nearly identical white-painted wooden crosses and picket fences.

San Diego County includes over 30 different California Indian cemeteries. Some date from the mission period, others from the formation of the reservations after 1875, and still others are only a few years old. California Indian cemeteries are Catholic cemeteries and are largely exclusive to individuals of indigenous descent. They still follow the *campo santo* tradition of the early Spanish Catholics, mirroring the chapel burial grounds of Spanish missions. A large cross traditionally marks the fenced tract of consecrated ground. The majority of grave markers are white wooden crosses, although there are some stone and cement grave markers as well.

Over 100 pioneer cemeteries in San Diego County were established by homesteaders in the late 19th and early 20th centuries. The term "pioneer" typically refers to Americans who moved West, even though Spanish and Mexican colonists were among the first of their culture to settle the region. Pioneer cemeteries, in terms of their layout and grave markers, are the most diverse type of burial ground in San Diego County. Typically the larger multi-family community graveyards evolved from these private pioneer cemeteries.

We have included over a dozen graveyards in the chapter "Large Cemeteries." These are of two types: community cemeteries and mega-cemeteries. Community cemeteries developed from local pioneer burial grounds. In most cases, an altruistic landowner or organization supplied an unused tract of land where neighboring families could bury their dead. These graveyards eventually became official town cemeteries. The second type of large cemetery, pre-planned mega-cemeteries, is strikingly different. While community cemeteries grew in a relatively unplanned manner, mega-cemeteries, holding up to 100,000 burials, were designed to accommodate huge numbers of interments from the start. Whereas community cemeteries evolve, modern cemeteries merely fill up.

We have attempted to make this an exhaustive text. However, there are private Native American cemeteries that we were asked not to include, specifically on Palomar Mountain and Mount Laguna and in the Santa Ysabel Indian Reservation. Furthermore, the SDGP is still tracking clues regarding the following graveyards and individual burials: the Cline Ranch Cemetery; the Baker Cemetery; the Fowler Ranch Cemetery; the Native American cemetery in Old Town; the graves of twins on Whale Mountain in Ballena; the burials on the old Sunset Ranch, now the Encinitas Golf Course; the de Larosa baby grave site in Jacumba; the Edwin Snow and Edward Foss grave sites in Alpine; the Mary Ward and Val Dukes grave sites in Ramona; the Mocogo Ranch burials in Dulzura; the Freeman burials in Moosa Canyon; James W. Robinson's Point Loma grave; the Gaskell child's grave site near Campo; the Augusta Ormsby 1871 grave in a cluster of century plants at Gopher Canyon and old Bonsall Road; Isaac Frazee's private cemetery where the Castle Creek Golf Course is now located; the Palomar Mountain graves, including that of Joseph Smith; and John Popplewell's 1935 Vista ranch grave.

In compiling the images and text for this book, we have repeatedly encountered evidence of gravestones being erased from the local landscape. Cemetery scholar Penny Coleman once noted, "People who create cemeteries expect them to last forever. However, they don't." Why does the phrase "grave desecration" evoke such a powerful emotional response and why should San Diegans be concerned about the legacy of lost graves? Ancient Roman law, as issued in Corpus Juris Civilis (529–534 A.D.) under the rule of Justinian I, decreed that "Every person makes the place that belongs to him a *religious* place at his own election, by *the carrying of his dead into it*" (italics added). It is the act of burial that makes a place sacred, and it is only through the protection of consecrated spaces and their markers that this legacy is upheld. Without cemeteries, we collectively suffer a loss of history, identity, and integrity. Furthermore, to erase burial markers and diminish memorialization is to belittle and corrupt our humanity.

One

Mission Cemeteries

Mission San Luis Rey De Francia, c. 1915. Founded in 1798, Mission San Luis Rey de Francia was one of the largest and most prosperous of the California missions. The third and present version of the mission was completed in 1815, secularized in 1834, and returned to the Catholic Church in a dilapidated condition in 1865. Restoration has been ongoing since the 1890s. (Courtesy Oceanside Historical Society.)

MISSION SAN LUIS REY CEMETERY, C. 1905. This cemetery was the site of a murder on January 13, 1863, when the infamous Cave Couts, owner of Rancho Guajome, dispatched his brother Blount to prevent the burial of smallpox victims. Blount Couts confronted the burial party of rancher Don Ysidro María Alvarado and shot Leon Vasquez in the face as the latter leapt onto the cemetery wall. Two others were wounded. (Courtesy Oceanside Historical Society.)

ENTRANCE TO MISSION SAN LUIS REY CEMETERY. The skull and crossbones over the gate are not the creation of mission friars. They are, in fact, a set decoration courtesy of Disney Studios. Several episodes of the 1950s television series *Zorro*, starring Guy Williams, were filmed at the mission. (Courtesy SCIC.)

Franciscan Burial Crypts. Just inside the gate to the Mission San Luis Rey cemetery is a stairway leading downward to the friars' burial crypt, the last resting place of many who served the mission since its establishment in 1798. The cemetery proper and the surrounding grounds hold the remains of Catholic clergy, pioneers, and almost 3,000 California Indians. (Courtesy SCIC.)

Asistencia de Las Flores, c. 1874. This outpost for Mission San Luis Rey provided support to travelers on El Camino Real into the 1840s. The adobe structure and corral pictured here were the site of a battle for the provincial governorship of Alta California between Jean Bautista Alvarado and Carlos Antonio Carrillo in 1838. The ruins are located within the present boundaries of Camp Pendleton Marine Corps Base. (Courtesy SDHS.)

ASISTENCIA DE LAS FLORES RUINS. Expanding into an estancia, Las Flores was transformed into an Indian pueblo during the Mexican period and then later absorbed into Rancho Santa Margarita y Las Flores. John Baumgartner, raised on the rancho in the early 1900s, stated in a 1982 oral history interview, "When I was a youngster, those walls were, oh, say ten feet high or more. . . . There was one or two graves there and it was fenced in. . . . as it rained, the adobe would melt." (Courtesy SDHS.)

ASISTENCIA DE LAS FLORES TODAY. All that remains of the *asistencia* is this small section of adobe wall; the rest is overgrown. There is no aboveground evidence of the graves. (Courtesy SCIC.)

Mission San Antonio de Pala, c. 1890. The Pala Mission was founded in 1816 as an *asistencia* to the larger Mission San Luis Rey de Francia. It was very prosperous until secularization by the Mexican government in 1846, when buildings began falling into disrepair. Partial restoration in 1903 led to a major reconstruction of the adobe structures in 1956. (Courtesy SDHS.)

Mission San Antonio de Pala Cemetery, c. 1900. Spanish Catholic funerary tradition viewed cemeteries (*campo santo*) as holy ground. Padres continued to use the space in and around their missions as cemeteries even if the structures themselves had fallen into ruin. The individually fenced graves seen here are characteristic of Catholic cemeteries of the period. (Courtesy Escondido History Center.)

Pala Mission Cemetery, c. 1960. This photograph was taken from the bell tower. It shows the cemetery after restoration and landscaping. A modern sign in the graveyard attests that this cemetery holds the remains of hundreds of indigenous converts, yet virtually none of the standing gravestones indicate that California Indians were buried within the walls of the cemetery proper. (Courtesy Escondido History Center.)

Mission Santa Ysabel "Brush Church." Santa Ysabel was an *asistencia* under Mission San Diego de Alcalá. It celebrated its first mass in 1818. By 1822, Mission Santa Ysabel had a chapel, cemetery, granary, and 450 neophytes. The decline of the mission system following secularization left the chapel in ruins, yet various accounts maintain that services have been held continuously since 1818, within interim structures such as this brush church, shown here about 1901. (Courtesy Museum of Man.)

MISSION SANTA YSABEL CEMETERY, C. 1897. This image shows the brush church and bells in the distance. George Tays described the area in 1937 as "a small cemetery, with a bare wooden cross and little mounds of sunbaked earth . . . guarded, by a tall picket fence, against profanation." (Courtesy Museum of Man.)

THE LOST BELLS OF SANTA YSABEL. Lorena Hoover rings the mission bells in 1883. They were purportedly the oldest mission bells in California, made in 1723 and 1767. The bells were stolen in 1926. While the clappers were found almost immediately and are on display at the mission today, the bells have yet to be recovered. (Courtesy Ramona Historical Society.)

SANTA YSABEL CHAPEL AND CEMETERY TODAY. The cemetery at Santa Ysabel has been used almost exclusively by California Indians since its establishment in 1818. Considering its small size, there are presumably many unmarked graves in the general vicinity. Santa Ysabel has no chapel burial ground dedicated to Euro-Americans, as is found at other missions. The present chapel, built in 1924, is the fourth to have been constructed on this site. (Courtesy SCIC.)

Two

California Indian Cemeteries

ANCIENT TRIBAL ETHICS PRESENT PROBLEM TO CITY IN MOVING GRAVES TO BUILD CAPITAN DAM

Indians Very Particular How Departed Spirits Are Handled to Avoid Suffering to Living; Two Old Cemeteries On Site; Real Action Delayed, but Natives Anxious.

Capitan Grande Indian Reservation Cemetery, Early 1930s. Prior to the 1930s, two bands of Diegueño Indians—the Los Coñejos and the Capitan Grande—had been burying their dead at the El Capitan Grande Indian Reservation Cemetery. In the 1930s, the two groups resettled to accommodate the flooding of a reservoir basin that was formed by El Capitan Dam. This resettlement included disinterring and transferring their dead. These pictures are the only known photographs of the original Capitan Grande Cemetery. The inset shows the grave of an unbaptized baby who could not rest in hallowed ground. (Courtesy *San Diego Union*.)

Viejas Chapel, c. 1957. The Los Coñejos resettled at the Viejas Indian Reservation. The chapel for the Viejas Indian Reservation, the Nativity of the Blessed Virgin Mary Catholic Church, stands next to the cemetery on Viejas Grade Road. (Courtesy Museum of Man.)

Viejas Indian Reservation Cemetery Today. The Viejas Indian Reservation Cemetery is still active and has existed since the Los Coñejos reburied their dead there in the 1930s. (Courtesy SCIC.)

BARONA INDIAN MISSION, C. 1960. Whereas the Los Coñejos band moved to Viejas, the Capitan Grande band relocated to the Barona Indian Reservation. In 1931, the Capitan Grande Indians—in direct competition with movie stars Douglas Fairbanks and Mary Pickford—purchased the Padre Barona Valley following the construction of the El Capitan Dam. Shown here is the reservation chapel, the Assumption of the Blessed Virgin Mary, on Wildcat Canyon Road. The cemetery is out of view to the right. (Courtesy San Diego State University Special Collections.)

BARONA INDIAN RESERVATION CEMETERY TODAY. The Lakeside Historical Society's *Legends of Lakeside* states, "The Barona Indians were first located in back of El Capitan Dam. I know several of the boys who helped move the graveyard from there to Barona Reservation." There were over 500 bodies in the Capitan Grande Cemetery. (Courtesy SCIC.)

RINCON CHAPEL, C. 1910. This image shows the Rincon Indian Reservation chapel when it was likely under construction. There is no bell and the doors appear to be missing. The chapel's architecture is nearly identical to that of the La Jolla Indian Reservation Chapel. The St. Bartholomew Cemetery is out of sight to the right at the base of the hill in the distance. (Courtesy Museum of Man.)

ST. BARTHOLOMEW INDIAN CEMETERY, C. 1910. The priest, located in the center of this picture under an umbrella, is likely blessing the cemetery for the Rincon Indian Reservation. (Courtesy Museum of Man.)

St. Bartholomew Cemetery Today. Note the trees growing from the two graves. The date of death for the grave on the left is 1916. (Courtesy SCIC.)

St. Bartholomew Gravestones. Large stone monuments are rare in Native American cemeteries. Several of the stones date from the early 1900s. The earliest marked gravestone in the cemetery is dated 1879. (Courtesy SCIC.)

PAUMA INDIAN CEMETERY, C. 1910. Local resident Eleanor Beemer described the cemetery in her 1934 book, *My Luiseno Neighbors*, with these observations: "More than a hundred mounds fill the space inside a fence, paintless picket with sagging barb wire. . . . Two graves are dignified by granite tombstones. Any others marked have low wooden crosses, some covered with faded black calico ripping loose where it was sewed on." (Courtesy Museum of Man.)

PAUMA GRAVES, C. 1920. Beemer further noted, "[But] no grave is so neglected as to lack dishware or trinkets, the possessions the dead no longer use, a china cup, an alarm clock, soup plates, odd-shaped medicine bottles, a Japanese teapot. All the glassware is amethyst, purple shades varying with the time of exposure to the sun, vases, glass coal lamps and candle sticks." (Courtesy SDHS.)

PAUMA GRAVES. The granite gravestones that Beemer referred to in *My Luiseno Neighbors* still exist. They are in the Ayal family plot in a corner of the Pauma Indian Cemetery. (Courtesy SCIC.)

PAUMA INDIAN CEMETERY TODAY. The Pauma Valley landscape has changed considerably over the last 100 years. The cemetery is now surrounded by orange groves. Once an austere plot with a few saplings, today it is almost hidden under an umbrella of trees. (Courtesy SCIC.)

Pala Indian Reservation Cemetery. A short distance to the east of Mission San Antonio de Pala, this Native American cemetery is likely the original burial ground for indigenous converts. The graveyard is still in use today. The structure under the American flag is a memorial to Pala Indians who served in the armed forces. It is decorated with a handmade metal eagle and helmets from past wars. (Courtesy SCIC.)

La Jolla Indian Reservation Chapel and Cemetery. Although out of focus, this early-20th-century photograph shows Catholic church dignitaries and well-dressed parishioners in front of the original chapel. The decoration on the door suggests that the image was taken on a day of religious significance; perhaps the photograph marks the occasion of the chapel dedication. The cemetery can be seen in the background to the right. (Courtesy Museum of Man.)

LA JOLLA INDIAN RESERVATION CEMETERY. Taken the same day as the previous image, it appears that the cemetery is being blessed in this photograph. The chapel is just out of view to the right. (Courtesy Museum of Man.)

LA JOLLA INDIAN RESERVATION CHAPEL AND CEMETERY TODAY. The chapel has been rebuilt and is used as a meeting hall; the cemetery is still in use. This is the official cemetery for the reservation. There are three others, the Cuca, Potrero, and Amago, which are on separate land holdings. These were later incorporated into the reservation. Photographs of these graveyards are not available for publication. (Courtesy SCIC.)

MESA GRANDE INDIAN RESERVATION CEMETERY. Located on the site of the traditional Native American village of Tukomuk, this reservation and its cemetery lie at the base of Angel Mountain on Mesa Grande. Still in use, the cemetery has been in existence since the late 1800s. The children seen here are participating in the twilight candle lighting ceremony of All Souls Day around 1938. (Courtesy SDHS.)

All Souls Day at Mesa Grande, c. 1906. Members of the reservation celebrate La Noche de las Velas (the Night of the Candles), the "commemoration of all the faithful departed." The ceremony is performed annually on November 2. The Chapel of St. Dominic is in the background. (Courtesy Museum of Man.)

Mesa Grande Cemetery Today. The cemetery has changed little in the last 100 years. The building in this photograph is the Chapel of St. Dominic. Built in 1898, it is no longer in use. (Courtesy SCIC.)

MESA CHIQUITA INDIAN CEMETERY. The plaque at the base of the large cross reads, "Mesa Chiquita Cementerio," but this burial ground is commonly referred to as the "Bloomdale Cemetery" for its location within the Bloomdale Ranch on Mesa Grande. (Courtesy SCIC.)

MESA CHIQUITA FUNERAL, 1938. Tragically military veteran Maximino Morelli (in the flag-draped coffin); his wife, Edith; and his mother-in-law, Barbara Ponchetti, died from botulism poisoning after eating home-canned fish. Their young son, Jim, survived. Out of sight to the left is a military honor guard. (Courtesy SCIC.)

KUMPOHUI INDIAN CEMETERY. Almost entirely forgotten, this small Mesa Grande cemetery is unofficially named after the perennial spring that flows past it. The age of the cemetery is unknown, but grave-marker burial dates average around the 1920s; the most recent is 1988. A caretaker replaces crosses as they decay, several of which can be seen leaning against a tree to the left. (Courtesy SCIC.)

KUMPOHUI GRAVE. Fresh flowers decorate an unnamed grave in the Kumpohui Indian Cemetery. (Courtesy SCIC.)

"Monkey Hill Cemetery," c. 1918. The construction of Lake Henshaw resulted in the inundation of the Native American village of San Jose in Santa Ysabel. The plan necessitated the removal of the village's protohistoric/historical cemetery, located at the base of Monkey Hill,

before the hill became an island. The human remains were boxed and transported two miles east to a new location. (Courtesy SDHS.)

SECOND MONKEY HILL CEMETERY, C. 1918. A second site was chosen for the reburial of Native American remains from Monkey Hill. The cemetery was moved far enough away to ensure safety from the rising waters of Lake Henshaw, while remaining within visual range of the original village site to the west. Boxes of skeletons can be seen on the left. (Courtesy SDHS.)

SECOND MONKEY HILL CEMETERY TODAY. In the 1980s, it was reported that cattle were grazing among the grave markers within the dilapidated fence. The present fence is solidly constructed to prevent disturbance, but there are no surface indications to identify the enclosure as a cemetery. The dead are all but forgotten. The low rise on the left is Monkey Hill. (Courtesy SCIC.)

Mataguay Indian Cemetery. With no historical documentation, this cemetery is unofficially named for its location on the Mataguay Scout Reservation in Santa Ysabel, once the Treanor Ranch. Overgrown and untended, its 16 grave markers are in poor repair. The cemetery is likely over 100 years old. It stands on an area just north of Matagual Creek and is overseen by the Santa Ysabel Indian Reservation. (Courtesy SCIC.)

St. Francis Chapel, Early 1900s. The original chapel was established in the 1830s for the Cupeños Indians on their traditional tribal land, Kupa, now Warner Springs. In 1903, the federal government displaced the tribe, forcing them to relocate to Pala on a "Trail of Tears." The chapel and cemetery were deeded back to the Native Americans in 1914 by the wealthy Oakland capitalist, William Griffith Henshaw. (Courtesy SDSU Special Collections.)

St. Francis Indian Cemetery, c. 1900. According to Ed Fletcher (1900–1996) of Warner's Ranch, there was a legendary Native American curse on the immediate area. He related, "So the Indians held a fiesta at Warner's the night before the Army was coming to take them, and pronounced a curse on the Warner's Ranch and all the past owners and future owners of the ranch. The fact is, they sacked and burned the church; they killed one of the priests, and it was really a bad time." (Courtesy Ramona Historical Society.)

ST. FRANCIS INDIAN CEMETERY, C. 1910. Cremation burials uncovered during archaeological excavations identified this site as a Cupeños burial ground; it was converted to a Catholic cemetery in the 1830s. (Courtesy SCIC.)

ST. FRANCIS INDIAN CEMETERY TODAY. Fletcher also mentioned that "it was about 1917 or '18, along in there, when the Indians held their first fiesta. Dad allowed them to come and bury their dead in the old cemetery. They held a three-day fiesta and I was there. At the end of the fiesta they had a powwow and they went into a dance and they removed the curse." (Courtesy SCIC.)

San Ysidro Indian Cemetery. One of two cemeteries within the Los Coyotes Indian Reservation, this historical burial ground is associated with the village of San Ysidro. Well maintained and still active, the cemetery saw use in the 19th century, before the reservation was established. (Courtesy SCIC.)

San Ignacio Indian Cemetery. Located on the middle fork of Borrego Palm Canyon deep within the Los Coyotes Indian Reservation, this austere cemetery is associated with the community of San Ignacio. (Courtesy SCIC.)

SAN FELIPE INDIAN CEMETERY. The "San Felipe Indian Cemetary," according to the hand-carved sign on the gate, is located on the eastern slope of Volcan Mountain. There is no historical documentation relating to the cemetery; likely it is over 100 years old. The graveyard is still in use by members of the Santa Ysabel Indian Reservation. (Courtesy SCIC.)

INAJA INDIAN RESERVATION CEMETERY, C. 1930. Members of the Inaja Indian Reservation place lighted candles around the graves as part of All Souls Day. All Souls Day immediately follows All Saints Day. According to Catholic tradition, the former is a day to celebrate the deceased who have attained the beatific vision in Heaven; the latter is a day to pray for those who have not. (Courtesy Museum of Man.)

INAJA INDIAN RESERVATION CEMETERY TODAY. Inaja is one of the smallest Native American reservations in San Diego County. Its cemetery is located in this idyllic valley southwest of Julian. (Courtesy SCIC.)

OLD CUYAPAIPE INDIAN CEMETERY. There are two cemeteries southeast of Mount Laguna on the Cuyapaipe Indian Reservation. The earlier of the two is pictured here. It was likely established in the 19th century. Archaeologists noted on a 1980 site form that the old Cuyapaipe cemetery included barbed-wire fencing and white crosses. This image shows the graveyard's recent renovations, as it is now surrounded by a chain-link fence. The burials are marked by black crosses. (Courtesy SCIC.)

NEW CUYAPAIPE INDIAN CEMETERY. A flood destroyed the Old Thing Valley Road and effectively blocked access to the original Cuyapaipe Cemetery. As a result, a new Cuyapaipe Indian Cemetery was created nearby. The large gravestone in the center of the picture is not a traditional burial marker; it is a commemorative stone that reads, "Our Ancestors . . . Cuyamaca To Cuyapaipe . . . August 9, 1983." (Courtesy SCIC.)

MANZANITA INDIAN RESERVATION CEMETERY. Native American graves are commonly decorated with flowers; some are freshly cut, while others are plastic or paper. These grave markers at the Manzanita Indian Reservation Cemetery have been decorated with bandanas, the corners attached to each arm of the cross. (Courtesy SCIC.)

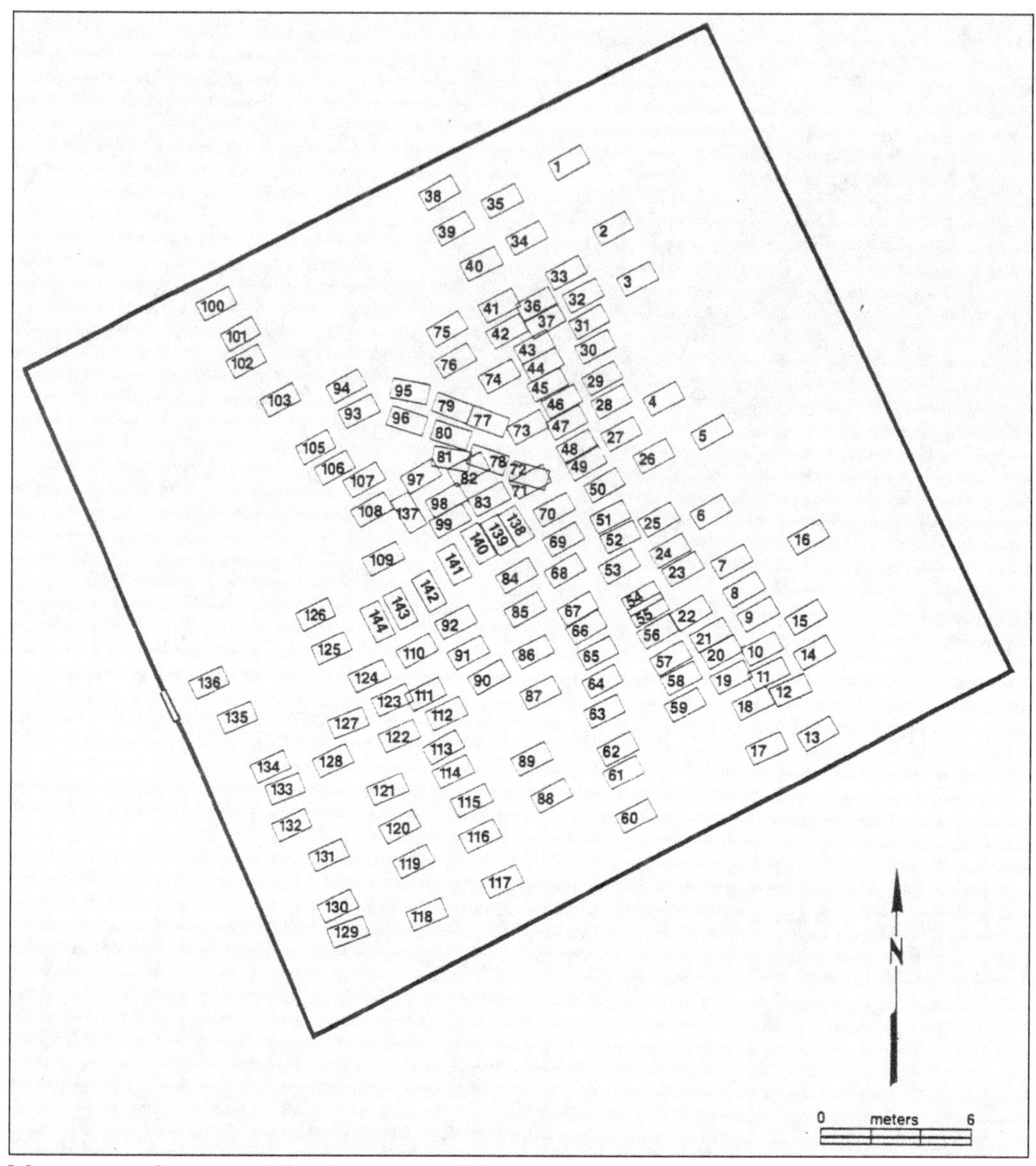

MANZANITA CEMETERY MAP. In 1980, WESTEC Services, Inc., mapped individual grave plots as part of an archaeological survey. They reported, "The cemetery consisted of 144 graves of which only 26 had names on the cross headstones . . . with some of the graves overlapping others." (Courtesy WESTEC Services, Inc.)

La Posta Indian Reservation Cemetery. There are three cemeteries on the La Posta Indian Reservation. The most recent of the three is pictured here. The earliest burial date is 1998. This burial ground stands on a bluff overlooking the earlier cemeteries two miles to the west. (Courtesy SCIC.)

Our Lady of Mount Carmel Chapel and Cemetery. There are two cemeteries on the Campo Indian Reservation. The one shown here is located next to the tribal offices on Church Road and is fronted by the reservation's chapel. The cemetery is still in use, but there is no information as to its age. The date of the photograph is unknown. (Courtesy SDHS.)

Campo Indian Reservation Cemetery, c. 1903. The second cemetery on the Campo Indian Reservation is just outside downtown Cameron Corners. In addition to the traditional crosses is a wooden tablet. (Courtesy Museum of Man.)

Campo Indian Reservation Cemetery Today. Over 100 years old, this tiny plot of ground is still in use and immaculately maintained. Not only are the crosses kept in pristine condition, the grave mounds are consistently reshaped to resist erosion. (Courtesy SCIC.)

JAMUL INDIAN VILLAGE, C. 1981. In 1912, three acres of hilly land were donated to the Roman Catholic Diocese to be used as a Native American cemetery. To fulfill the stipulation, the church built a small chapel and buried hundreds in the consecrated ground. Over the next 60 years, various Native American families moved onto the unoccupied areas of the cemetery and built their homes on territory that was originally reserved for the deceased. (Courtesy El Cajon Historical Society.)

ST. FRANCIS XAVIER CHAPEL. The Roman Catholic Diocese of Monterey erected this chapel for the Jamul Indian Village around 1912. It is rarely used now. A larger church has been constructed nearby. The Jamul Indian Village Cemetery is to the right of the chapel. (Courtesy SCIC.)

JAMUL INDIAN VILLAGE CEMETERY. According to pioneer Ella McCain, a Mrs. Maxfield "gave the land for the Indians' cemetery and it has been used ever since as a burial place for Indians in a radius of about seventy-five miles." The contrasting black and white crosses are unique to this cemetery. (Courtesy SCIC.)

SYCUAN INDIAN RESERVATION CEMETERY. There is no known written documentation concerning the Sycuan Indian Reservation Cemetery, but it has been in use since the establishment of the reservation in 1875. (Courtesy SCIC.)

Three

Pioneer Cemeteries

Ellis Ranch Cemetery. This pioneer cemetery is located in Descanso at the intersection of Highway 8 and Highway 79 North. It is still in use by descendants of Charles Ellis, who inherited the land from his in-laws, the Aguilar family. Charles's father-in-law, Gavino Aguilar, had operated the Santa Gertrude Ranch until his murder in 1882. The Ellis family owned and operated the Ellis Store and the Ellis Ranch Resort, as well as caring for the family cemetery. (Courtesy SCIC.)

AGUILAR GRAVESTONE. With questionable spelling, the brass plaque reads, "Grandfather Agalar—Ambushed & Killed by Andronico Lopez 21. Year Old—Murdered Plot To Get Agalar Land. Ambush was 1. Mile South Of This Grave—This Stone Came From Ambush Site. Andronico Lopez Died In SanQuinten Prison. Marker Erected By Frank–Tina & Dorra Ellis May 30." (Courtesy SCIC.)

ELLIS CEMETERY COMMEMORATIVE MARKER. This monument was dedicated in 1971 to commemorate the 100th anniversary of the arrival of Charles Ellis to the Descanso area, where he met and married Ysidora Aguilar. (Courtesy SCIC.)

SWAIN FAMILY CEMETERY. The Pines Fire swept through part of Descanso in 2002. Barely visible in the ash are the six known gravestones of the Swain Family Cemetery, located just south of the ranger station. Pioneers Samuel (1848–1908) and Lavina (1852–1945) Swain operated the Oak Grove Hotel. Their son-in-law and daughter, William and Edith Berkey, ran the hotel's Shamrock Saloon and the Oak Grove Store. (Courtesy SCIC.)

SWAIN GRAVESTONE. In the face of a wildfire, even gravestones are perishable. Lavina Swain's marker is partially buried in ash and debris. Cracking and spalling are evident around the edges of the granite flush marker. Thermal stresses have caused the raised letters to fall off as if they were just decals. Without restoration, this gravestone will soon be unidentifiable. (Courtesy SCIC.)

LOUISA WILLETT, DATE UNKNOWN. Born in Maine in 1813, Louisa Willett migrated to Alpine in 1883. She named her homestead Mount Pisgah, after the biblical mount from which Moses viewed the Promised Land. Disturbed that families had to either bury their dead on their property or transport them to downtown San Diego, Willett set aside a portion of her land for a cemetery. (Courtesy Alpine Historical Society.)

MOUNT PISGAH CEMETERY, 2003. Fire and vandals long ago destroyed the wooden markers of the 14 known pioneers buried behind the water tank in Alpine Heights. The cemetery was in use from 1890 to cemetery founder Louisa Willett's death in 1907. Clemens Granite Works donated the memorial stone in 1994. (Courtesy SCIC.)

WILLIAM E. FLINN, C. 1864. At a time when transportation was limited, private ranch cemeteries were used by multiple families within a neighborhood. It was through the generosity of family patriarch William Flinn (1817–1896) that the private Flinn Springs Cemetery became the burial ground for the local area. This graveyard was active from around 1876 through the 1950s. (Courtesy El Cajon Historical Society.)

FLINN SPRINGS CEMETERY TODAY. The Flinn Ranch was subdivided long ago. Development has encroached to the point that the cemetery is now located within a mobile home park. Highway 8 is just out of view to the right. Nevertheless, the landowners have preserved the graveyard as historically accurate as possible. The large granite gravestone to the left marks the graves of William E. Flinn and his wife, Sarah. (Courtesy SCIC.)

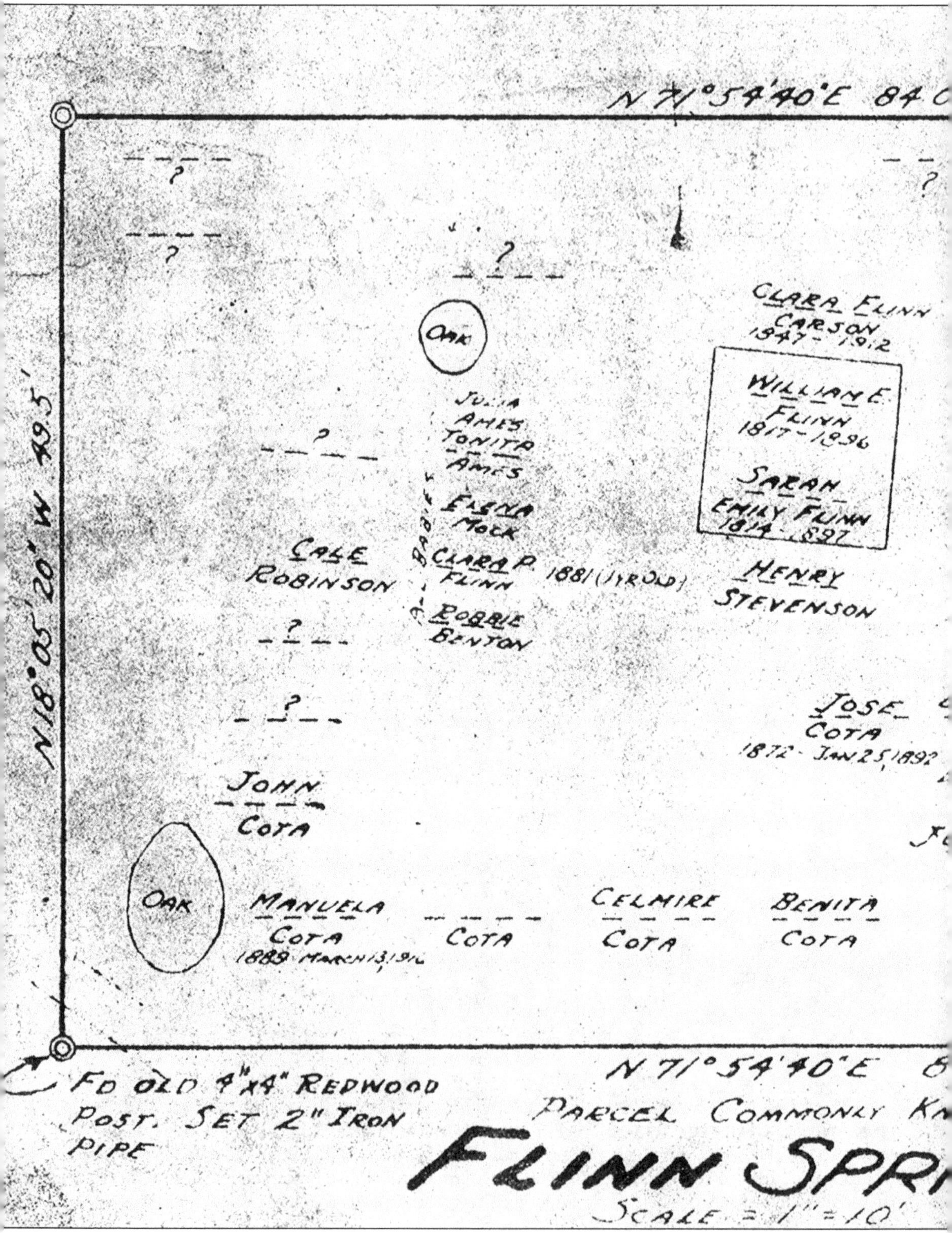

N 71°54'40"E 84…

N18°05'20"W 49.5'

?

?

?

Oak

Clara Flinn
Carson
1847 - 1912

William E.
Flinn
1817 - 1896

Sarah
Emily Flinn
1814 - 1897

Julia
Ames
Tonita
Ames

?

Elena
Mock

Babies

Cale
Robinson

Clara P.
Flinn
1881 (1yr old)

Henry
Stevenson

Robbie
Benton

?

?

Jose
Cota
1872 - Jan 25, 1892

John
Cota

Oak

Manuela
Cota
1889 - March 13, 1916

Cota

Celmire
Cota

Benita
Cota

N 71°54'40"E 8…

Fd old 4"x4" Redwood
Post. Set 2" Iron
Pipe

Parcel Commonly Kn…

Flinn Spri…

Scale = 1" = 10'

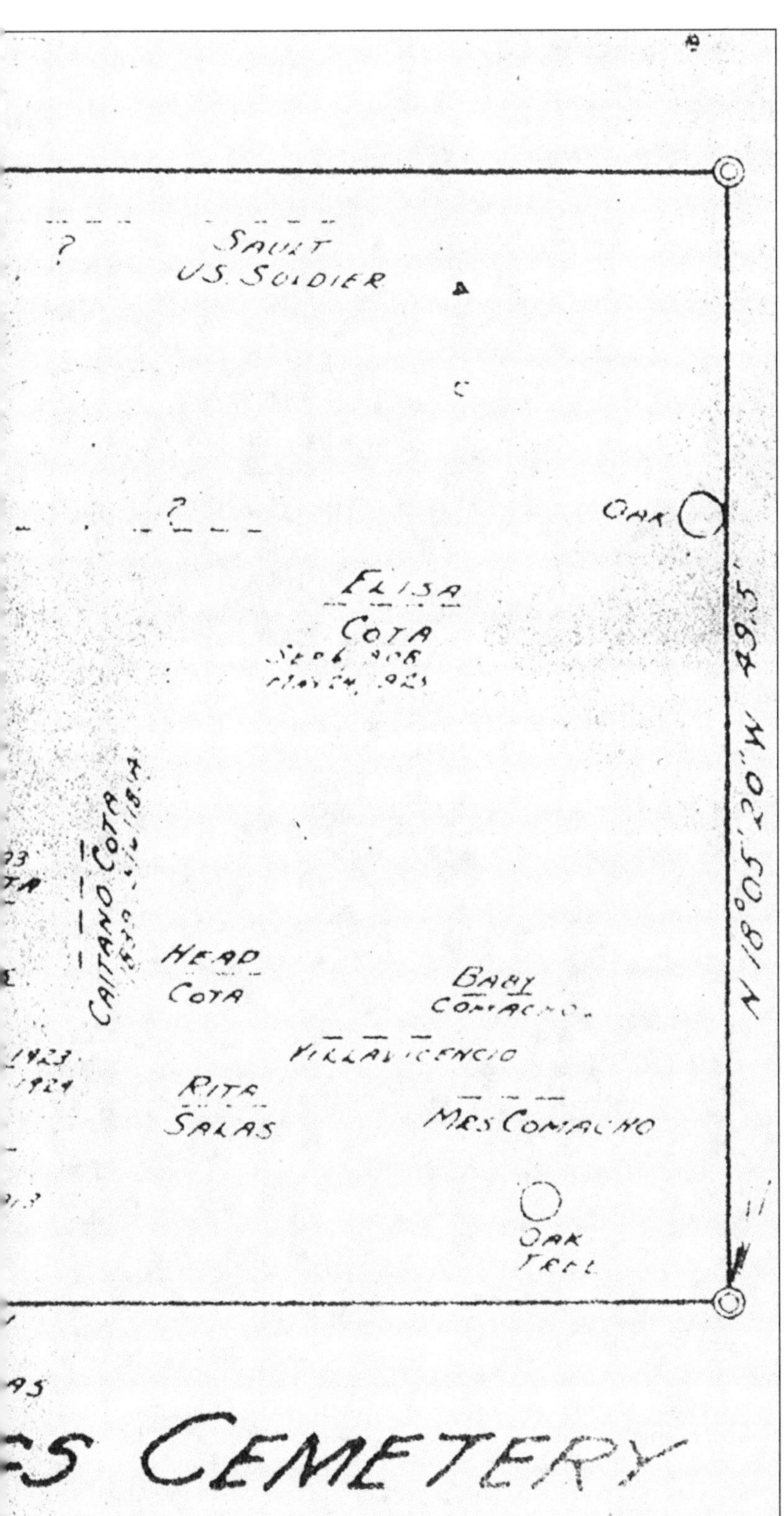

Flinn Springs Cemetery Map, Date Unknown. Descendant Julia Flinn De Frate described her family's cemetery in 1953. She noted that "Little Elena Mock was buried beneath a big oak. . . . Eventually, at least fifty people who had lived and died near the Flinn Ranch found their final resting place in the little cemetery. With passing time and changing ownership of the land, the Flinn Springs cemetery was almost forgotten . . . there are still graves that cannot be identified." (Courtesy SCIC.)

MONUMENT HILL. The marble plaque on this large rock reads, "In memory of Stephen Brayton. Born March 17, 1797. Died January 12, 1884." Once sheltered by a pepper tree, this boulder off Monument Hill Road in El Cajon stands over pioneers Stephen Brayton, Thomas L. Treat (1795–1887), and an unknown adult and child. The land was once the Brayton homestead. (Courtesy SCIC.)

WHITAKER GRAVES. Author David Caterino and a student assistant record the gravestones covering the cremated remains of longtime Lakeside residents Hale (d. 1980) and Mildred (d. 1992) Whitaker and Mildred's mother, Nellie Scharnke (d. 1955). The Whitakers hand-built their hilltop home of native stone between 1935 and 1940. The County of San Diego is preserving their property at the end of Castle Court Drive. (Courtesy John Gibbins/*San Diego Union-Tribune*.)

Muth Baby Girl Grave Site, 1986. This low-quality image is the only photographic record of the restored grave of Baby Muth (d. 1900), daughter of A. M. Muth, for whom Muth Valley was named. Archaeologists initiated a preservation plan after discovering that the grave had been desecrated and the marker removed. Unfortunately, the site was destroyed by fire approximately 15 years later. (Courtesy SCIC.)

Smith Cemetery. This cemetery is all that remains of the Caskey-Winfree-Smith-Campbell group that arrived in Poway from Kentucky around 1884. These pioneers constructed a plantation-style house that they named Longview. Rev. Michal Smith, for whom the cemetery is informally named, purchased the abutting property, on which the cemetery is located. Garden Road can be seen in the distance. (Courtesy SCIC.)

POWAY PIONEERS. A communal gravestone honors those buried in the Smith Cemetery, now devoid of individual markers. The last to be buried was Reverend Smith's sister, Nora Scroggs, who died at the end of the 1800s. Longview burned during the winter of 1907–1908, and both properties were eventually sold. The graduated rod is a measuring tool used by the San Diego Gravestone Project. (Courtesy SCIC.)

FOSTER GRAVE SITE, DATE UNKNOWN. William Frederick Foster (1824–1881) was a major landowner in the San Dieguito Valley, farming much of what is now the Del Mar Fairgrounds. He arrived in the area around 1850 and was buried on his property. Foster's gravestone was still standing into the 1960s on high ground just east of Highway 5. It has since disappeared. (Courtesy Del Mar Historical Society.)

ANGIER CEMETERY DESTROYED. Albert W. Angier's neighborhood cemetery stood on what is now a parking lot behind these buildings on Via de la Valle, across from the Del Mar Racetrack. There were eight known burials, dating from 1886 to 1919, including Angier and his wife. A 1928 legal action by an heir to declare the land a public cemetery failed; only a few bodies are known to have been exhumed. (Courtesy SCIC.)

OLIVENHAIN CEMETERY. The German settlers who developed the north county neighborhood of Olivenhain established this burial ground, originally owned by the colony's first mailman, Louis Denk. The first recorded burial was that of Amelia Hauck, who died in 1891 at age 29. The cemetery is still active, although burial is limited to descendants of the original pioneer families of Olivenhain. (Courtesy SCIC.)

MEADOWLARK RANCH CEMETERY, 2003. Two weathered crosses, the largest for Tomasa Elisa Gonzalez Peralta de Tico (1867–1906), were all that remained of the Tico family's cemetery in San Marcos when road widening began on Rancho Santa Fe Road at Meadowlark Ranch Road. The concrete pad seen through the fencing protects the five unmarked graves. (Courtesy SCIC.)

SAN MARCOS PIONEER. Tomasa Elisa Gonzalez Peralta de Tico and her second husband, Juan Tico, homesteaded 160 acres along San Marcos Creek. Tomasa's brother, Salvador Gonzales, deeded the land for the cemetery. Their great-grandfather, Jose Francisco de Ortega, was a Spanish soldier and pathfinder for Fr. Junipero Serra, founder of the first of San Diego's missions. (Courtesy San Marcos Historical Society.)

MEADOWLARK RANCH CEMETERY TODAY. The cemetery was severely disturbed by utility crews in 1959. Then the Tico family fought developers for years as attempts were made to exhume the bodies. The family won their emotional and legal battles, and the cemetery survives, protected under tons of concrete in the center of Rancho Santa Fe Road. (Courtesy SCIC.)

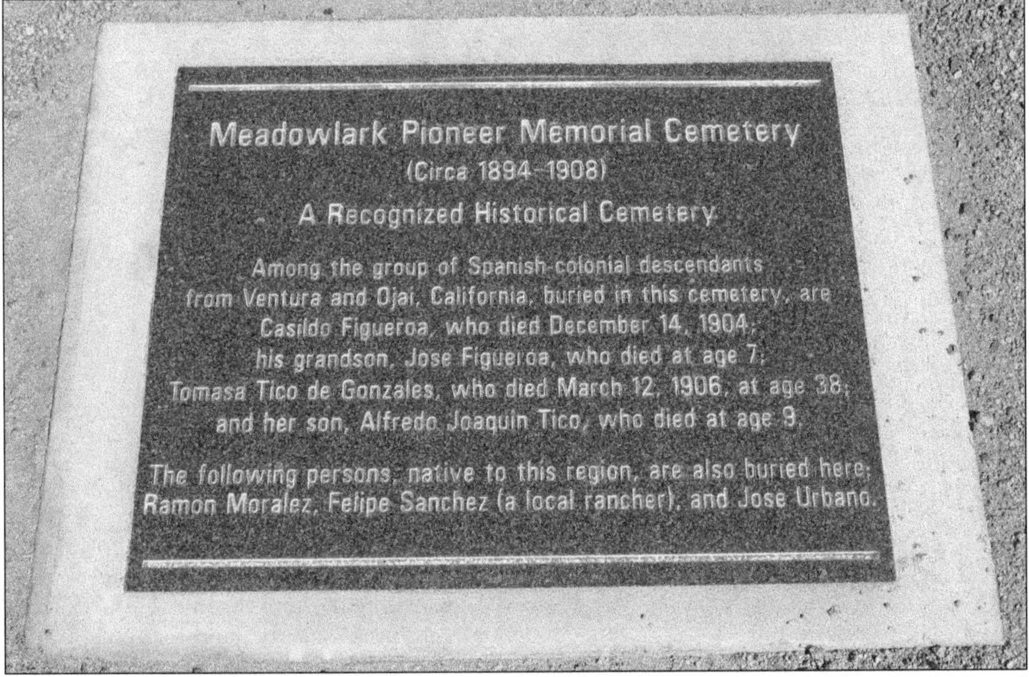

CLOSE-UP OF MEADOWLARK MEMORIAL. The marker pictured here rests over the Tico family's burial ground. Ramon Moralez is one of the individuals buried in the Tico Family/Meadowlark Ranch Cemetery. He was killed by Parker Dear Jr., grandson of pioneer Cave Couts. (Courtesy SCIC.)

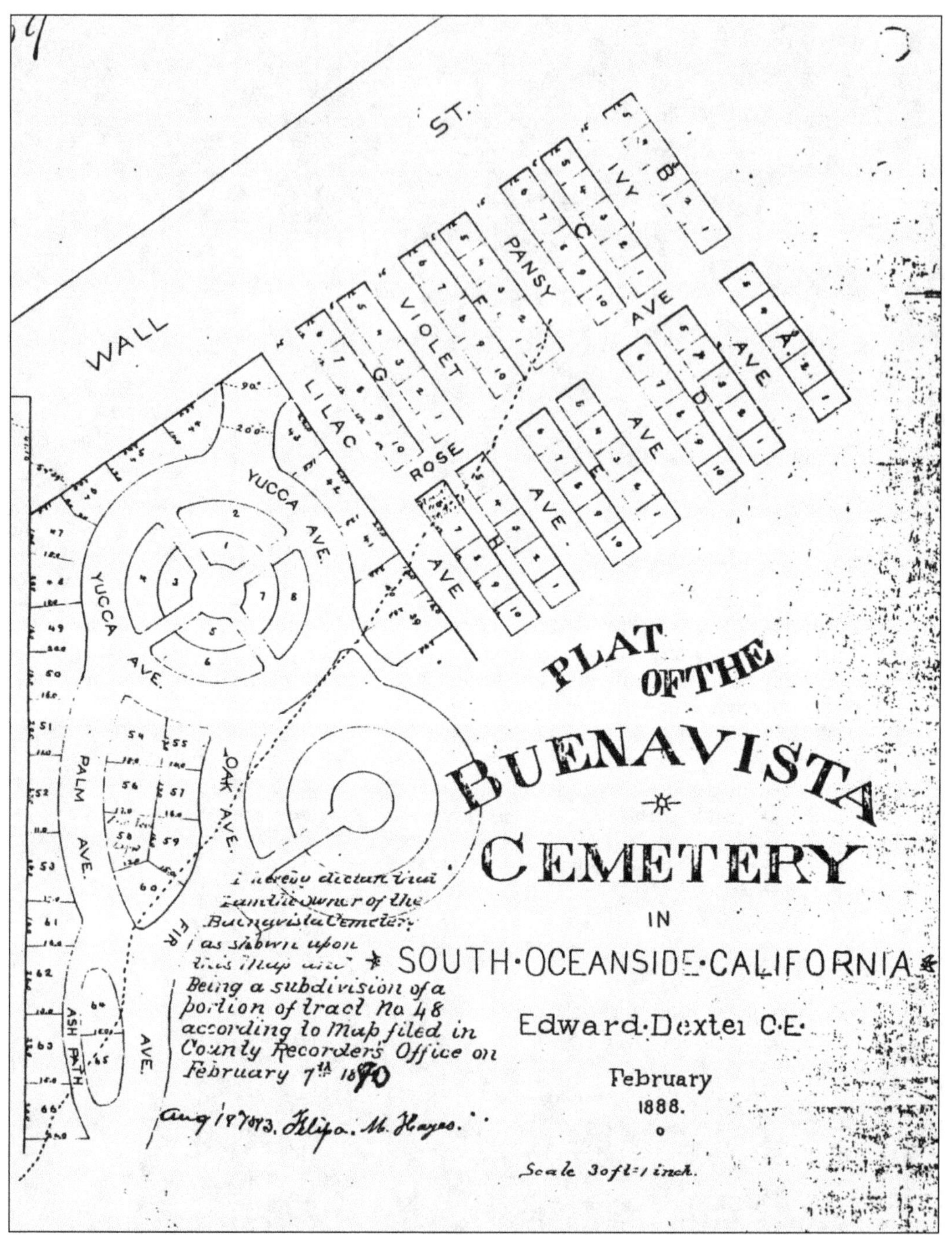

Buena Vista Cemetery, c. 1890. This map for the Buena Vista Cemetery on Wall Street—now Vista Way—was officially recorded in 1893 by Oceanside's real estate magnate, J. Chauncey Hayes. Hayes started the cemetery as a commercial enterprise, but the land had been in use as a local graveyard since the 1880s. The last burial was around 1906. (Courtesy Oceanside Historical Society.)

BUENA VISTA CEMETERY, 1966. A 1952 fire burned the fence and all wooden grave markers, effectively erasing the location of many burials in the Buena Vista Cemetery. A 1964 *San Diego Tribune* article called it the "Mystery Cemetery." This rare image of the graveyard was taken during a beautification project. The cemetery was erased by development several years later, with construction workers unearthing bodies and quickly reburying them. (Courtesy SDHS.)

"HAUNTED RESTAURANT." Author Seth Mallios poses before the Hunter Steak House that was built atop the Buena Vista Cemetery in 1970. The sidewalk plaque at the bottom of the photograph lists 45 individuals buried in the cemetery. Unfortunately, plans to exhume the bodies prior to construction were only casually followed. There have been numerous reports of alleged paranormal activity within the restaurant. (Courtesy SCIC.)

San Luis Rey Pioneer Cemetery. This cemetery was part of the old town of San Luis Rey (1848–1888) and holds many of its earliest residents. Locals began referring to it as the "Pioneer Cemetery" to distinguish it from the cemetery at Mission San Luis Rey de Francia, seen in the distance. The first recorded burial was that of Catherine E. Foss in 1869. Burials have continued to the present, with a peak between 1880 and 1933. (Courtesy SCIC.)

S. D. Lanpher Gravestone. This exquisite marble marker stands proudly in San Luis Rey Pioneer Cemetery. The tree stump represents a life cut short. It is one of the few gravestones found undamaged on San Luis Rey's "rattlesnake hill." Unfortunately, it was once common practice for vandals to chain gravestones to their automobiles and drag them away. (Courtesy SCIC.)

ALFORD A. FREEMAN, DATE UNKNOWN. A transplant from Texas, A. A. Freeman (1822–1898) was the patriarch of the Freeman family. At one time, the San Luis Rey Pioneer Cemetery was known as the "Freeman Cemetery" because of the prominent number of Freeman graves. (Courtesy Ilene Rufrano.)

ALFORD A. FREEMAN GRAVESTONE. Hand-carved gravestones are very rare in San Diego County. Sandstone was selected for this marker, a type of native rock easily worked with hand tools. It reads "IN MEMORY OF A. A. FREEMAN—BORN APRIL 29, 1822—DIED NOV. 29, 1898." Crude representations of palm fronds can be faintly seen on the upper corners. (Courtesy SCIC.)

ST. MICHAEL'S-BY-THE SEA EPISCOPAL CHURCH. Established in 1894, Carlsbad's first church of record was moved several blocks north from Oak Avenue and Lincoln Street. Old St. Michael's has been preserved as a historic landmark next to a larger sanctuary. Its small cemetery was destroyed. The bodies were exhumed—the exact number unknown—with several reburied at the All Saints Episcopal Church Cemetery. (Courtesy *Blade-Tribune* 1959.)

ALL SAINTS EPISCOPAL CHURCH, C. 1920. Located on Peyri Road behind Mission San Luis Rey, this small, Gothic-style church catered to the spiritual needs of the English colony that settled San Luis Rey Valley between 1888 and 1890. The church was established in 1890 with funds provided by the Church of St. Savior's, Ealing, England. (Courtesy Oceanside Historical Society.)

All Saints Episcopal Cemetery, c. 1980. Bishop Ford Nichols consecrated the church and cemetery in 1892. By 1945, a steadily dwindling congregation necessitated an association be formed to maintain both the church and burial ground. Services are now held only twice a year, on Memorial Day and All Saints Day. Burials are limited to church members and those deemed qualified by the Episcopal Diocese. (Courtesy Edith Swaim.)

All Saints Grave Site. Francis W. Reynolds (1865–1896) was the first person to be buried at All Saints Episcopal Cemetery. His daughter, Mary Reynolds, was the first to be christened in the church, in 1893. (Courtesy SCIC.)

RANCHO GUAJOME, C. 1895. There are believed to be three graves in the immediate vicinity of the Rancho Guajome chapel, seen here on the left. Two are children, one from a servant and the other the owners' child. An obituary in the November 3, 1855, *San Diego Herald* stated, "DIED. On Saturday, Oct. 27th, at Rancho Guajomito, very suddenly, ABEL, only son of Col. CAVE J. COUTS, of Guajomito, aged 3 years and 9 months." (Courtesy SDHS.)

RANCHO GUAJOME CHAPEL GRAVESTONE. This handmade marker for Pedro Munoz is partially buried in a flower bed. He is assumed to be a ranch hand, but it is unknown if the inscribed date, October 19, 1925, is that of his birth or death. (Courtesy SCIC.)

Fox-White Cemetery. Active since the burial of Fred E. Fox in 1904, this hilltop site off Live Oak Park Road is the only recorded private cemetery in Fallbrook. Still active, its last recorded burial was in 1999. The cemetery is just uphill from what is believed to be the oldest standing residence in Fallbrook. (Courtesy SCIC.)

Stewart Family, c. 1901. Fox, Clark, King, Stewart, and White are just a few of the Fallbrook names found in the Fox-White Cemetery. The gravestones of Fred (1876–1950) and Ellen (1884–1950) Stewart, pictured here, can be seen in the foreground of the previous photograph. Their daughter, Gertrude (Stewart) Morse—sitting on her father's lap in this image—died and was buried in the Fox-White Cemetery in 1962. (Courtesy Fallbrook Historical Society.)

Kolb Family Cemetery. Scattered across this hilltop north of Rainbow are 25 gravestones, fragments, and plot markers dating from 1885 to 1979. The exact number of burials is unknown. The stone home of the Kolb family, built in 1920, is out of view to the left. Surprisingly, this cemetery, located on the Riverside County border, is still active. According to the present owner, an elderly Kolb descendant is planning to be buried there. (Courtesy SCIC.)

San Pasqual Cemetery. One of the most picturesque cemeteries in San Diego County, this hilltop burial ground known as "Cemetery Knoll," "Boot Hill," or "Cemetery Ridge," is the final resting place of many local pioneers and their descendants. A cemetery association was officially formed in 1911, with land donated by the Trussell and Rockwood families. It is located on Highway 78 at Santa Ysabel Creek. (Courtesy SCIC.)

HIGGINS FAMILY CEMETERY. This graveyard is located east of Couser Canyon Road. Researchers named this the Higgins Family Cemetery based on the nearby James P. Higgins farmstead and the extant headstones of Higgins's wife, Nancy (d. 1907), and Margery Ann Higgins (1849–1875). Ground-penetrating radar surveys hint at other interments in the graveyard. Written accounts suggest that an unnamed baby and Mrs. Tom Rogers are buried there. The Higgins Family Cemetery may have been a community burial ground. (Courtesy SCIC.)

MISSING GRAVES. Grave markers for members of the Dye and Mulkins families were last seen on the hillside behind the ranch buildings in the 1960s, east of Ramona near Sutherland Dam Road. Archaeological investigations have yet to locate the graves of these individuals, buried between 1860 and 1880. The Dye property later became the Rotanzi Ranch. (Courtesy SCIC.)

LITTLEPAGE CEMETERY. A Ballena resident poses within the overgrown cemetery, reportedly established by Martin Casner upon the death of his son, Martin, in 1875. An elderly resident revealed that her uncle, the caretaker, had accidentally started a brush fire years ago, destroying the wooden grave markers. More recently, the Littlepage Cemetery was used as a horse corral. The gravestones date from the 1870s to the 1970s, many of them identifying Littlepage family members. (Courtesy SCIC.)

LOST MARCKS GRAVE SITE. In 1887, beekeeper William Marcks of Germany homesteaded this site overlooking Mildred Falls, seven miles south of Santa Ysabel. Marcks died in 1931 and was survived by a wife and four sons. The property was ravaged by brush fires in 1957 and 2003, erasing all evidence of the reported grave of one of Marcks's sons. The foundation seen here belonged to the honey extraction house. (Courtesy SCIC.)

Eagle Peak Cemetery. A mysterious hilltop cemetery is located in the Cleveland National Forest just southwest of Wynola. There is no official name for the cemetery, and there are no longer grave markers to identify the occupants. One possible hint, from a 1949 historical note, reads, "Mr. & Mrs. Walton . . . buried on ranch . . . Eagle Peak." (Courtesy SCIC.)

A Missing Cemetery. The backyard of 4701 Mountain Brook Road in Wynola was once the Reed Ranch Cemetery; the grave markers disappeared during development of the area in the 1970s. At least nine people are known to have been buried around 1880–1910, including owners Robert and Antonia Reed, several family members, and Wynola locals. (Courtesy SCIC.)

Bailey Ranch Cemetery. Wynola's Bailey Ranch is now the Canta Rana Ranch, but the cemetery still exists deep in the hills. Records list eight burials: the original ranch owner, Charles Bailey, and seven members of the Bailey and Brawley families. (Courtesy SCIC.)

Bailey Gravestone. Only one gravestone remains in the Bailey Ranch Cemetery, that of Mrs. James (Sue) Bailey (1837–1882). The grave marker is likely a replacement stone that was placed in the 1970s. (Courtesy SCIC.)

Scholder Family, Date Unknown. German immigrant Frederick Scholder (1831–1917) married a California Indian woman, Joaquina Saubel Navalet (d. 1931). Scholder, known historically in the area as "The Squaw Man," was the first non–Native American settler on Mesa Grande, where he and his wife raised 14 children. Shown here are Frederick Scholder (center), Joaquina (with dog), and three of their children. (Courtesy Museum of Man.)

Scholder Family Cemetery. This hilltop enclosure surrounds at least 10 graves, with dates ranging from 1904 to 1978. The centerpiece of the cemetery is the gravestone of Frederick and Joaquina Scholder. Their impressive monument is surprisingly elaborate for a rural cemetery. It is carved from red granite, a material not native to the area. (Courtesy SCIC.)

ANGEL FAMILY, DATE UNKNOWN. James and Henrietta Angel settled their Mesa Grande ranch in 1880. Over the years, they raised nine boys, two girls, and thousands of head of cattle. A nearby peak, Angel Mountain, bears their name. (Courtesy Ramona Historical Society.)

ANGEL FAMILY CEMETERY. Although Angel descendants still live on the original family property at the base of Angel Mountain, their family cemetery is no longer in use. The total number of burials is unknown, but there are at least five: James Newton Angel (d. 1934); his wife, Henrietta (d. 1939); Woodson Angel; John Angel; and Mark Angel. (Courtesy SCIC.)

John D. Stone, c. 1900. John was the patriarch of the Stone family, a well-known name on Mesa Grande. The Stone, Morris, and Ambler families intermarried and were closely associated with the Mesa Grande Store. Arthur Stone worked there for 30 years; his brother-in-law, Cleason Ambler, was postmaster in the store for four decades. (Courtesy Museum of Man.)

Stone Family Cemetery. This small plot is hidden deep in the woods on Mesa Grande, on what was once Stone property. John D. Stone (1841–1910) is buried in the center, with wife, Lute (1848–1928), and son James (1884–1963), from left to right, respectively. (Courtesy SCIC.)

WATKINS-ALFORD CEMETERY. Mesa Grande rancher Spike Alford stands next to the gravestone of his great-grandparents, Morgan R. Watkins (1846–1913) and Mary C. B. Watkins (1853–1914). Included in the overgrown plot are several unmarked graves, including one belonging to Spike's great-uncle Arthur Watkins. A tiny headstone marks the grave of a neighbor's baby, Gregory Steel. Until recently, this cemetery was almost completely unknown outside of the family. (Courtesy SCIC.)

ED AND MARION DAVIS, C. 1909. Mesa Grande's Ed Davis was famous for his Powam Lodge, built on the Cherry Hill Ranch. He was an authority on Native American lore and accumulated an extensive collection of artifacts. Davis traveled extensively with his camera and is responsible for many of the photographs of San Diego pioneers and California Indians. (Courtesy Museum of Man.)

DAVIS MEMORIAL. The centerpiece of the Davis Cemetery, this enormous boulder houses the cremated remains of Edward Davis (1862–1951) and his wife, Anna Marion Davis (1862–1932). In the foreground is a viewing platform built over the steep hillside. The cemetery proper is behind the boulder. It has been in use since 1913, with the last burial in 2002. (Courtesy SCIC.)

DAVIS FAMILY CEMETERY. The Davis Cemetery is almost covered with ornamental plants. In the foreground is the gravestone of Bob and Jo Davis. Behind are the footstones, engraved with the initials "KND" and "HLD," of Kathryn N. Davis (1889–1916) and Harvey L. Davis (1886–1913), respectively. Their monuments are out of sight to the left, carved in rock. (Courtesy SCIC.)

Angelo and Julia Botti, c. 1890. Angelo Botti (1872–1936) emigrated from Italy with a sherbet recipe. He met his Czechoslovakian wife, Julia (1876–1966), in New York. They owned an ice cream and candy establishment in Los Angeles, moving to Santa Ysabel in 1920 upon discovering the curative effects of the Warner Hot Springs on Angelo's arthritis. (Courtesy Taylor family.)

Botti Family Cemetery. Within this shady grove, Angelo and Julia Botti are buried together under a concrete ledger. Their grandson, Charles J. "Ange" Botti Jr. (1926–1996), is interred next to them. The Taylors, descendants of the Bottis, still live on the property at the base of Mesa Grande. (Courtesy SCIC.)

TREANOR GRAVE SITE. Deep within the Mataguay Scout Reservation between Santa Ysabel and Warner Springs, a large boulder marks the grave of the previous landowner, John Treanor (1883–1935), and the cremated remains of his wife, Catherine Elizabeth Treanor (1887–1976). The monument lies within a one-acre parcel retained by the Treanor family as stipulated in the deed to the reservation. (Courtesy SCIC.)

JACK BILLINGSLY GRAVE SITE. The rectangular hole carved into this boulder on Inspiration Point at Warner Springs once held cremated remains. A bronze plaque removed by vandals read "Jack Billingsly 1885–1938." Billingsly was reported to have worked at Warner Hot Springs in the 1920s and 1930s. (Courtesy SCIC.)

EAGLE'S NEST GRAVE SITE. Before being patented for the Los Coyotes Indian Reservation in 1913, a mountaintop tract was homesteaded by Hiram Keyes and Eugene Pannenberg in the late 1880s. The land was then used by the Fletcher family as a retreat called Eagle's Nest Ranch. The lonely grave shown here, outlined by a post-and-chain fence, likely belongs to one of these families. (Courtesy SCIC.)

THE HELM FAMILY, C. 1898. The Helms were one of the first families to settle in the Warner Springs area. Their descendants can still be found in the region today. This photograph contains intricate details. Note the girls on the left sitting sidesaddle. In addition, the woman on the right appears to be blind. A close-up of this photograph revealed that she also has a large scar that runs from her lip to the back of her neck. (Courtesy SDHS.)

HELM FAMILY CEMETERY. The remains of this pioneer burial ground sit atop a knoll just outside the Los Coyotes Indian Reservation. Time and an abortive attempt to restore the cemetery have all but obliterated the grave markers. Reportedly buried in the Helm Family Cemetery are Turner Helm, Wid Helm, a pair of twins, and several unknown individuals. (Courtesy SCIC.)

KIMBLE-WILSON STORE "BOOT HILL." The adobe Kimble-Wilson Store in the distance, now in ruins, was a local meeting place for the Warner Springs area. A burial ground was established on the hill just a stone's throw away. The alleged murderer of Joseph Smith of Palomar Mountain was hung here in 1868. Edward Davis noted in his 1938 book, *Palomar Mountain History*, that the suspect "was taken to a live oak tree at the foot of the hills convenient to the store, a noose was fitted around his neck . . . and 25 men heaving, they pulled him up . . . and left him." (Courtesy SCIC.)

Harry Moore Forster Gravestone. In 1935, there was only one gravestone behind the Kimble-Wilson Store. It belonged to Harry Moore Forster, who died in 1888 at the age of 57. At least three others are believed buried there: 1) Daunt Helm; 2) the British deserter who murdered Joseph Smith of Palomar—then Smith—Mountain; and 3) an anonymous man of such considerable size that an extra-large casket had to be manufactured and transported from Warner Springs for him. (Courtesy SDHS.)

John "Daunt" Helm Gravestone. Currently there is still only one gravestone at the site: a replacement marker for Daunt Helm. It is mislabeled—it reads "John Helm 1873"—and is mistakenly placed on the broken base of Harry Moore Forster's gravestone. Although legend states that a false grave was used to divert suspicion away from Daunt Helm in a murder case, he was in fact buried here after he died in his 70s. (Courtesy SCIjC.)

WILSON FAMILY CEMETERY. The Wilson, Paroli, and McManama families are represented in this small burial ground on the west side of San Felipe Valley. Handmade grave markers, such as these made of sheet metal, identify 20 graves. (Courtesy SCIC.)

PAROLI FAMILY CEMETERY. The Paroli family maintained several homesteads in the San Felipe Valley during the early 1900s while supervising the San Felipe rancho. Their cemetery is located on a slope overlooking San Felipe Valley. The handmade decorated cross, pictured here, is dated 1934; modern gravestones date from the 1970s to the 1990s. (Courtesy SCIC.)

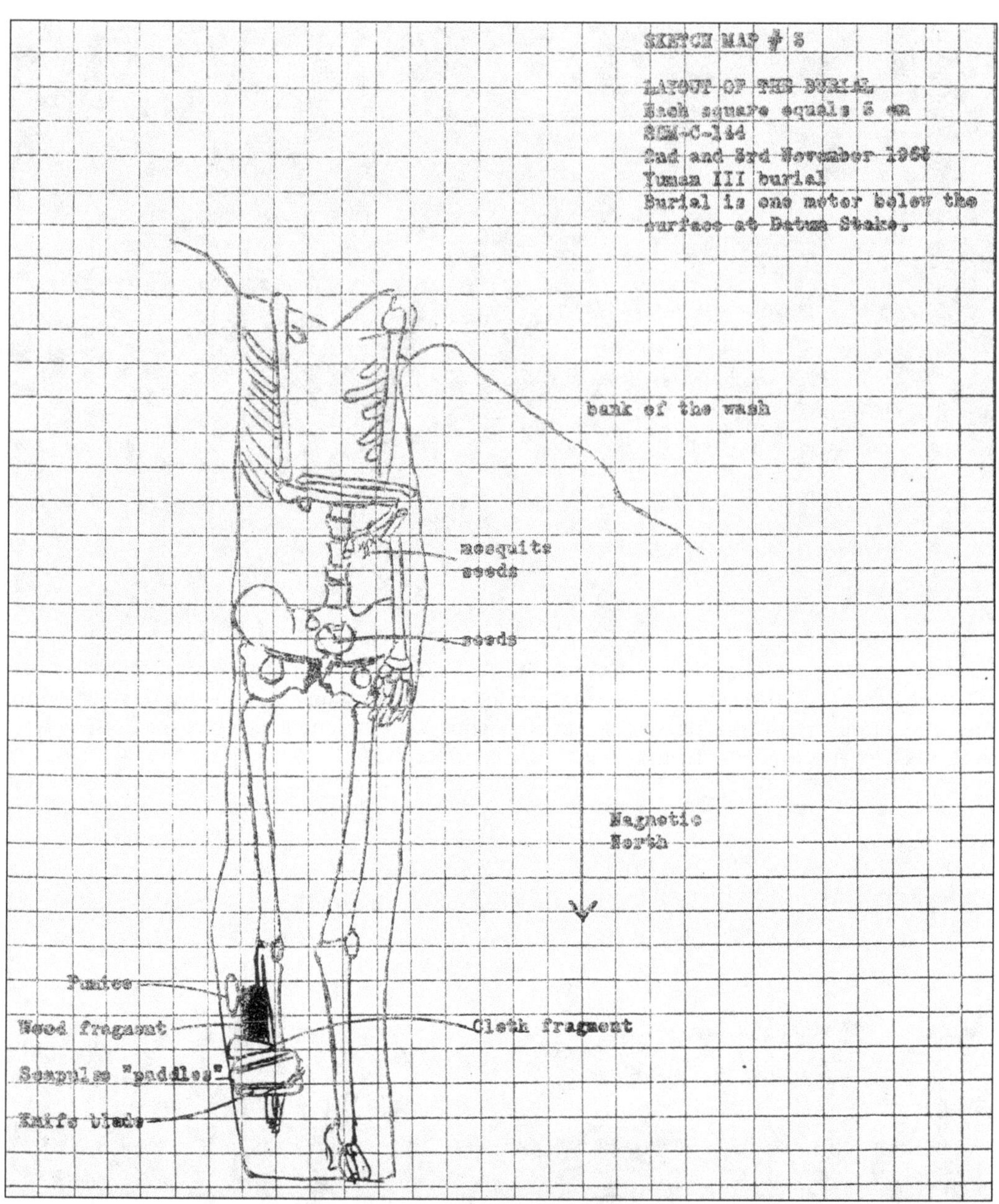

Mason Valley Archaeological Excavation. In 1963, a skeleton was discovered eroding from the side of a wash near the old Mason homestead, located in Mason Valley north of Campbell Grade. Salvage archaeologists removed the remains, assuming them to be Native American. Mason Valley had been a Kumeyaay village prior to 1879. However, since it was the Native Americans' practice to bury cremated remains, there is the possibility that the skeleton was that of Jesusa Mason. (Courtesy SCIC.)

MASON HOMESTEAD, C. 1901. In the center of this photograph are rancher James Mason and his wife, Jesusa. The other individuals are believed to be son Edward and daughter Clara or Jessie. Jesusa died around 1905, followed later by a daughter; both were buried nearby. Floodwaters have destroyed much of the landscape, including the grave sites. James Mason's ashes are buried at Vallecito Stage Station. (Courtesy SDHS.)

VALLECITO STAGE STATION, C. 1910. Built in 1852, this adobe stage station was an important stop for the "Jackass" and Butterfield Stage Lines and for emigrant caravans. Several ghosts reportedly haunt the site, the most famous being the Lady in White. A Butterfield Stage passenger, this ill-fated woman sickened and died in 1869. Historian James Jasper reported that she was buried in "all her bridal toggery." The famed Lady in White has allegedly been seen parading about the station in full regalia. (Courtesy County of San Diego.)

JOHN HART GRAVESTONE, C. 1923. John Hart's grave site is shown as it looked over 80 years ago, at the Vallecito Stage Station. In 1866, Hart purchased squatter's rights to the station that he and his wife ran until he died a year later, at age 30. He was purportedly instrumental in the ambush of three horse thieves within the station. Their bodies are buried somewhere on the grounds. (Courtesy County of San Diego.)

VALLECITO CEMETERY. The present cemetery is a reconstruction for tourists. The cairn holds James Mason's ashes, although the date of death is incorrect (he died in 1929, not 1931). John Hart's 1867 white marble tablet with a willow tree motif—pictured in the previous image and on the far right in this photograph—is one of the oldest gravestones in San Diego County. (Courtesy SCIC.)

TOWN OF BANNER, C. 1891. Time has not been kind to Banner, seen here against Banner Grade. A flood all but obliterated the town in 1926. The cemetery, on a shelf against the side of Volcan Mountain, narrowly escaped. Unfortunately, a brush fire in the 1930s destroyed all but the marble gravestone of L. G. A. Coursolles (d. 1873), "A Native of Canada East." Now that, too, is gone. The cemetery is out of sight to the right. (Courtesy SDHS.)

"WHISKEY ROW." The Banner Cemetery had a notorious reputation. Ida Wellington noted in 1949 that among those buried there were "a number of drunkards who 'died with their boots on.' A number of Indians, graves unmarked. No trace of many. Several unknown miners killed in the mine." This grassy shelf still holds many pioneer graves. The "B" for Banner on the mountainside is a local landmark. (Courtesy SCIC.)

STONEWALL MINE, C. 1900. Cuyamaca City was a prosperous 19th-century town located south of Julian that faded into obscurity with the collapse of the Stonewall Mine. In 1981, archaeologist Daniel Foster recorded local Granville Martin's recollection of the cemetery associated with the Stonewall Mine. Martin noted that "there is parts of the old picket fences and stuff and I think there was parts of the boards or something that had been headstones. . . . there must have been five or six to seven graves." (Courtesy SDHS.)

STONEWALL MINE CEMETERY. Park ranger Pat Valenta stands atop Cuyamaca City's cemetery. Buried here were Jerome MacDonald and a Mr. White, both crushed by logs at the mine and eventually exhumed and moved. A third grave held the remains of Joseph McLane (d. 1889), who quit his job at the mine and went prospecting. The July 12, 1889, issue of the *Julian Sentinel* reported that McLane was found dead surrounded by "two dozen empty whiskey bottles and demijohns [large liquor bottles]." (Courtesy SCIC.)

Chilwell-Campbell Grave Site. William Chilwell was buried on his Laguna ranch in 1888 after being crushed by his horse. His wife, Louisa, married his business partner, Archibald Campbell; both were later buried at the site. Their epitaph reads, "If you seek their monument look around." The site overlooks Laguna Meadow in the Laguna Mountain Recreation Area. (Courtesy SCIC.)

Archibald and Louisa Campbell Grave. There are six known burials at the Chilwell-Campbell grave site: William Chilwell, Archibald and Louisa Campbell, Nell C. Kemp, Flora Chilwell Money, and Trevor Kemp. The dates of death range from 1888 to 1989. At some time in the past, all of the gravestones were collected and placed within Archibald and Louisa's grave enclosure. (Courtesy SCIC.)

AMOS BUCKMAN, DATE UNKNOWN. In 1876, Buckman discovered the Buckman Springs, or Soda Springs, as they were locally known, and promoted the healing properties of his Buckman Springs Lithia Water. The endeavor was never successful because the water was cloudy and discolored. The remains of his bottling plant, store, and home can still be seen off Olde Highway 80. (Courtesy SDHS.)

AMOS BUCKMAN GRAVE SITE. Buckman was buried on his property in the town that bears his name. The fence that borders Highway 8 makes a jog around his gravestone. The inscription reads, "Amos Buckman—Died Mar. 28, 1898 Aged 77 Years & 18 Days.—Located The Buckman Springs In 1876." (Courtesy SCIC.)

Cameron Corners Cemetery, c. 1929. Sometimes referred to as the "Campo Cemetery" or "Boot Hill," this tiny hillock was the local burial ground for the area from Campo to Jacumba. It was in use at least as early as the 1880s. (Courtesy SDHS.)

McCain Gravestone. Much of San Diego County was still dangerous frontier country into the early 20th century. This gravestone from the Cameron Corners Cemetery reads, "William McCain was killed by the Indians at Jacomba Feb. 27, 1880. Aged 17 Yrs. 7mos. & 7ds." (Courtesy SCIC.)

Box Tomb. Traditionally made of stone, box tombs are rare in San Diego County. This is the only known local example of a brick box tomb, occupying a plot in the Cameron Corners Cemetery. Brick would have been the obvious construction material in a place where a round-trip to San Diego for granite, on horseback, could take over a week. The occupant is unidentified. (Courtesy SCIC.)

Cameron Corners Cemetery Today. At least 22 individuals have been buried on "Boot Hill," their markers long gone. Unfortunately, vandalism is not a recent phenomenon. Ella McCain, whose ancestors lie in the cemetery, wrote in 1955, "Vandals recently went into the little cemetery, moved the headstones and broke them to pieces. Someone placed the beautifully carved white headstones together, and one can still read: William McCain, killed by Indians." (Courtesy SCIC.)

WAT GARNER GRAVE SITE. Little is left of Ancel Watson "Wat" Garner's hillside grave site, destroyed by time and fire. It is halfway up the hill, in the center of the photograph. Garner died on his ranch in Hayden Valley around 1895, at age 32, and is buried just inside the Manzanita Indian Reservation border next to his daughter Stella. His wife, Sallie, may also be next to them. (Courtesy SCIC.)

ORTEGA FAMILY CEMETERY. The Kemp family now owns the former Ortega property, with an easement granted to the family cemetery. The majority of burials are from the 1920s to the 1950s, although the cemetery is still in use. Cameron Corners can be seen in the distance; the tall building is the Motor Transport Museum of San Diego. Also on Kemp property, the Cameron Corners Cemetery is among the trees in the distance to the far right. (Courtesy SCIC.)

CAMP LOCKETT, C. 1947. Established as a border post in 1941, this U.S. Army reservation and POW camp was a cavalry camp since 1878. Two burial sites pre-date the reservation. Buried just inside the present camp gate were three bandits who attempted to rob the old Campo Store and were lynched for their misdeeds. At some point, their graves were washed into the creek. However, the second burial site, a pioneer cemetery, still exists. (Courtesy SCIC.)

CAMP LOCKETT PIONEER CEMETERY TODAY. Located behind the sheriff's station, this small pioneer burial ground was referred to as the Campo Cemetery. There are no longer any gravestones to identify the occupants, but an inventory from 1949 gives a partial listing: Allister (?) Gaskell, Honora Gaskell, Mrs. Beckley, "and some others." (Courtesy SCIC.)

"This Place Is Good Enough for Me." So said pioneer Capt. Charles G. McAlmond (1827–1887) of Potrero Valley. He brought his family to the area in 1868 and built an adobe house in 1871. McAlmond and his wife, Alpha (1837–1912), were postmaster and assistant postmaster of Potrero, respectively. They also operated a stage line from San Diego to Campo. The fence in the foreground marks the site of the McAlmond Family Cemetery. (Courtesy SCIC.)

McAlmond Gravestone. The individual graves of the McAlmond family are no longer distinguishable. At some point in the past, a communal gravestone was placed on their graves in Long Potrero and surrounded by a protective chain-link enclosure. Peter Jenson is believed to have been a laborer. (Courtesy SCIC.)

POTRERO CEMETERY. Potrero can thank John J. Williams for its cemetery. He deeded the little plot of land to the valley when his daughter, Alameda Pearson, died in 1887. In the foreground are the gravestones of two pioneers, Damon Thing (d. 1905) and his son Joseph (d. 1940). The name "Thing" is well known in San Diego; Thing Valley, Thing Road, and the Thing Brothers Store in Tecate. (Courtesy SCIC.)

WALKER RANCH CEMETERY. In 1896, Indiana natives Albert and Sarah Walker established a ranch in Deerhorn Valley (Jamul), now the Bradford Ranch. That family has maintained the cemetery since 1985. The plot was once distant from the original homestead but is now in the center of the ranch complex. Six individuals are believed to occupy the plot, including Sarah Walker (d. 1901) and her daughter, Laurel. The cross and statue are recent additions. (Courtesy SCIC.)

Four

Large Cemeteries

Alpine Cemetery, c. 1952. B. R. Arnold donated the land that is now the Alpine Cemetery in 1899. That same year, Sara Long was the graveyard's first burial. The Alpine Cemetery Association was formed in 1902, the articles stating, "The Alpine Cemetery now contains the graves of more than six human beings who have been interred therein." (Courtesy Alpine Historical Society.)

ALPINE CEMETERY FUNERAL, 1932. This photograph of the burial of Mary Ann Cuthbert (1872–1932) provides a rare view of the early Alpine Cemetery. For more than 50 years, graves were dug by hand, with preparation of the graves being a family responsibility. Recently, burial space has been restricted to Alpine residents to ease a shortage of plots in the graveyard. Prior to this restriction, 75 percent of the interments were for non-residents who were taking advantage of the inexpensive plots. (Courtesy Alpine Historical Society.)

CUTHBERT GRAVESTONE. The granite marker of Mary Ann and John Cuthbert from the previous photograph was displaced, though not destroyed. (Courtesy SCIC.)

GERADEHAND GRAVE MARKER. A 1929 fire destroyed all but a few of the early wooden grave markers in Alpine Cemetery. The open areas in the older portion of the cemetery are comprised of those burial plots with no identification. One of the few markers to be spared was that of Frank Adolph Geradehand (1845–1900). (Courtesy SCIC.)

ALPINE CEMETERY TODAY. The tradition of informality and hands-on family involvement required of early Alpine burials continues today. It is common to see graves outlined with handmade borders and decorated with folk art. The plot borders must be removable to allow for the passage of the grave digger's backhoe, a nod to modern technology. (Courtesy SCIC.)

SINGING HILLS MEMORIAL PARK, PRE-1996. Singing Hills is the most recent cemetery to be established in San Diego County, the first in almost 40 years. The land selected on Dehesa Road in El Cajon was used for agriculture from the 1920s to the 1940s, then mostly abandoned. In order to maintain the history and flavor of the valley, the 35-acre cemetery is surrounded by a wildlife preserve of almost 65 acres that is accessed by hiking trails. (Courtesy Singing Hills.)

SINGING HILLS MEMORIAL PARK, C. 1996. This composite photograph was taken as construction neared completion. The facility is designed to hold 30,000 burials and 20,000 entombments. It is expected to meet the funerary needs of the area for at least 30 years. (Courtesy Singing Hills.)

El Cajon Cemetery. Before the 1902–1903 establishment of El Cajon Cemetery, the city's first graveyard, residents either had to transport their dead to downtown San Diego or bury them in private graves nearer to home. In fact, prior to its official designation as El Cajon Cemetery, the immediate area had been a burial ground for years; the oldest recorded tombstone bears the date 1889. In addition, unmarked graves have been found outside of the present cemetery bounds. (Courtesy SCIC.)

"Dry Section." The oldest section of the El Cajon Cemetery is easily identified by the upright markers, scattered trees, and poor ground cover. The combination of sandy soil, shallow graves, and lack of burial vaults has made it necessary to avoid watering the grass because of the inherent risk of ground collapse. Consequently, this historical area of the cemetery has been deemed "the dry section." In 1969, the graveyard was expanded with a flush marker–only policy. (Courtesy SCIC.)

HALL GRAVESTONE. The oldest tombstone in the El Cajon Cemetery is that of Mariah and John Hall, who died in 1889 and 1891, respectively. The Hall family was so large that El Cajon's "Once-A-Week" newspaper quipped, "There is a certain house near the center of the Valley of but ten rooms which contains twelve Halls." The Halls helped develop and maintain the cemetery for generations. (Courtesy SCIC.)

BASCOM FAMILY PLOT. The Bascoms are buried in the oldest section of the El Cajon Cemetery. D. S. Bascom owned the El Cajon Meat Market. (Courtesy SCIC.)

Glenn Abbey Memorial Park, c. 1926. An October 1, 1926, *San Diego Business* article proclaimed that Glenn Abbey Memorial Park is "surely one of the most attractive portals in San Diego, the great wide gates opening on acre upon acre of verdant lawn. . . . Here much has been done to remove the gloom and ugliness from man's last resting place." Bonita's Glenn Abbey was founded in 1924 as the first endowment-care memorial park in the Southwest. The cemetery did not allow upright gravestones. (Courtesy SDHS.)

Little Chapel of the Roses, c. 1926. Glenn Abbey's most notable element is the non-sectarian Little Chapel of the Roses. It is a replica of Alfred, Lord Tennyson's chapel in Somersby, England. The circular stained-glass window is illustrative of his poem, "Crossing the Bar," a common theme inscribed on many of the grave markers at Glenn Abbey. The chapel is as popular for weddings as it is for funerals. (Courtesy SDHS.)

Glenn Abbey Memorial Park and Mortuary Today. Even though Glenn Abbey was at the forefront of progressive cemetery development strategies, it initially maintained discriminatory practices. A deed granted to a Harriet M. Lodge in 1928 reads, "That said land shall be held for cemetery purposes only, and shall be used for underground burial of human dead of the white race and none other." (Courtesy SCIC.)

Kathy Fiscus (1945–1949). Three-year-old Kathy Fiscus fell into an open well in San Marino, California. Volunteers worked frantically for over two days to free her, only to find that she had died. Thousands watched transfixed as the event unfolded, experiencing live, on-site television news broadcasting for the first time. Soon after, "Kathy Fiscus laws" were enacted requiring the capping of abandoned wells. Fiscus is buried at Glenn Abbey. (Courtesy SCIC.)

FRANK KIMBALL, DATE UNKNOWN. Builder Frank Kimball and his brothers Warren and Levi, the founders of National City, purchased 26,632 acres of land in 1868. Thirty-three acres from this Rancho de la Nacion were set aside for a cemetery on a mesa overlooking Sweetwater Valley and San Diego Bay. The first recorded interment was in 1870. (Courtesy National City Public Library.)

FRIDAY, FEBRUARY 11, 1870.

Raining a little in A.M. With Warren walked out to Sweetwater and laid out Cemetery and set men to work cutting out road and digging grave for Mrs. Wincapaw –

Foom & Sutter 2d
Lumber for Cart $10.00

KIMBALL DIARY. Frank Kimball recorded the first day at the La Vista Cemetery, writing, "Raining a little in a.m. With Warren walked out to Sweetwater and laid out Cemetery and set men to work cutting out road and digging grave for Mrs. Wincapaw." Frances Wincapaw was a native of Maine; her grave is unmarked. (Courtesy National City Public Library.)

GILSTRAP TOMB. The box tomb of William Harvey Gilstrap (1857–1933) is one of the most elaborate memorials in La Vista Memorial Park. His cryptic epitaph—in both English and Spanish—reads, "Versed In Cosmos And The Industrial System. A Friend To The Poor and Producing Class." (Courtesy SCIC.)

LA VISTA MEMORIAL PARK TODAY. Over time, La Vista fell into neglect and decay. Today it is a perpetual-care, lawn park facility where flush markers are mandatory. However, "Rest Haven," Kimball's original La Vista Cemetery, is still maintained much as it was, with upright gravestones. As a non-endowment area, only general maintenance is performed; family members must maintain individual graves. (Courtesy SCIC.)

DR. LOUIS N. HILLEARY, DATE UNKNOWN. The land for Poway's Dearborn Cemetery was donated in 1885 by Dr. Louis Hilleary (1854–1907), the first doctor in the area. Hilleary arranged funerals, acted as mortician, supplied the team of horses for the hearse, and drew up the original plot map. (Courtesy Poway Historical Society.)

HILLEARY GRAVESTONE. Dr. Louis Hilleary is buried in the oldest section of the cemetery he founded, now known as Dearborn Memorial Park. At some time in the past, his grave marker was pushed over and has been left lying horizontally to discourage further vandalism. (Courtesy SCIC.)

Dearborn Gravestone, Date Unknown. Although John T. Dearborn (1841–1886) was the first to be buried in the cemetery that was named after him, it includes gravestones that predate his marker. Originally, a man by the name of Thomas Rhodes had established a small burial ground on land he homesteaded. However, he aroused the local ire by refusing to accept further burials and announcing his attention to plow the land. Thirteen bodies were transferred from the Rhodes ranch to Dearborn Cemetery. (Courtesy Poway Historical Society.)

Dearborn Memorial Park Today. In 1963, expansion and modern improvements were begun on the cemetery, including ground leveling. Leslie Franz reported in the *Sentinel Poway Weekender* (May 21–27, 1977), "Somehow a pile of dirt left above a grave was scary to many people. . . . It gave them a feeling that a body was lying there." The catalpa trees, planted by Hilleary when the land was a timber claim, still remain. (Courtesy SCIC.)

OAK HILL MEMORIAL PARK. The first recorded burial at this Escondido cemetery was that of a young child, Lena Abbie Hays, in 1878. The Oak Hill Cemetery Association was officially founded in 1889. Ramon Montiel cleared the land with his horses, patterning the cemetery grounds in the shape of a spoked wheel. The decorative arch was built by local blacksmith and Oak Hill occupant Albert Bandy (1902–1975). (Courtesy SCIC.)

A GARDEN CEMETERY. The expansive Oak Hill Memorial Park exemplifies the rural cemetery movement in the north county. This movement replaced the previously harsh imagery of death with elements of art, gardening, architecture, and city planning. In this park-like setting, the Catholic and Mennonite churches, the Masons, Odd Fellows, and the Grand Army of the Republic held title to separate sections. (Courtesy SCIC.)

Escondido's First Librarian. During a 2007 event, Lindsey O'Connor portrays Mina Ward (1875–1902) at Oak Hill Memorial Park. The grave marker is a hollow metal casting of a zinc-bronze alloy. Now quite rare, such white-bronze monuments were marketed at the end of the 19th century as an inexpensive alternative to marble or granite. (Courtesy SCIC.)

Sophie Cubbison's Gravestone. Born Sophie Huchting in 1890, Cubbison received a degree in home economics from California Polytechnic State University in 1912. Once she substituted her famous melba toast for bread crumbs in stuffing, Mrs. Cubbison's All-Purpose Dressing became a household staple. It has lined grocery shelves since 1948. She was buried at Oak Hill in 1982. (Courtesy SCIC.)

San Marcos Cemetery. The cemetery was created in 1894 when the Littlefield family buried its matriarch, Nellie. Because the isolated town lacked a cemetery, the Littlefields and other area families formed the San Marcos Cemetery Association. Members maintained their own areas of responsibility within the original seven-acre site. Many improvements were made during the 1930s as part of the Works Progress Administration (WPA). (Courtesy SCIC.)

Adam Gale Malloy (1828–1911). Born in Ireland, Malloy served in the Civil War as commander of the 17th Wisconsin Volunteer Infantry. He was brevetted to brigadier general, U.S. Volunteers for gallant and meritorious service. When the U.S. Army was reorganized in 1866, Malloy accepted the rank of second lieutenant in order to remain in the service. These two gravestones, located in the San Marcos Cemetery, mark his grave and his career; one names him as a colonel, the other as a general. (Courtesy SCIC.)

IOOF Seal. The Independent Order of Odd Fellows is a fraternal organization. Inscribed on their seal is the imperative instruction, "BURY THE DEAD." With that tenet in mind, the Oceanside Odd Fellows Cemetery Association established the Oceanview Cemetery in 1895, where the common man could be assured a decent final resting place. (Courtesy SCIC.)

Oceanview Memorial Park. Commonly referred to as Oceanview Cemetery, the non-denominational, non-endowment facility was sold to the Ocean View Memorial Park Corporation in 1932, changing the financial focus to profit making. A string of financial setbacks forced the cemetery to eventually be sold to the Eternal Hills Cemetery Association in 1952, which still performs basic maintenance. (Courtesy SCIC.)

OCEANVIEW MAUSOLEUM THEN AND NOW. In 1932, the cemetery was expanded for a planned chapel and mausoleum. The top photograph shows Oceanview's nine-tiered edifice in 1970. Due to the cemetery's financial woes, construction of the structure was never completed, and it became a home to transients. Only a fraction of the 336 crypts were filled. Most of the structure was razed in 1985; the gated alcove was walled off, forming the bunker-like structure seen today. In poor, leaking condition and rarely unlocked, the mausoleum is reminiscent of a Hollywood horror movie set. (Right courtesy SDHS; below courtesy SCIC.)

Eternal Hills Memorial Park. Oceanside's largest cemetery was established in 1949 as a full-service, non-denominational, endowment care facility. Spread over approximately 165 acres, some undeveloped, it includes thousands of burials and entombments. Although flush markers are the norm, there are some upright gravestones. Eternal Hills also maintains the Oceanview Cemetery. (Courtesy SCIC.)

Jewish Cemetery. Eternal Hills is divided into multiple sections specific to religious affiliations, fraternal organizations, and so on. One such section is the North County Jewish Cemetery. (Courtesy SCIC.)

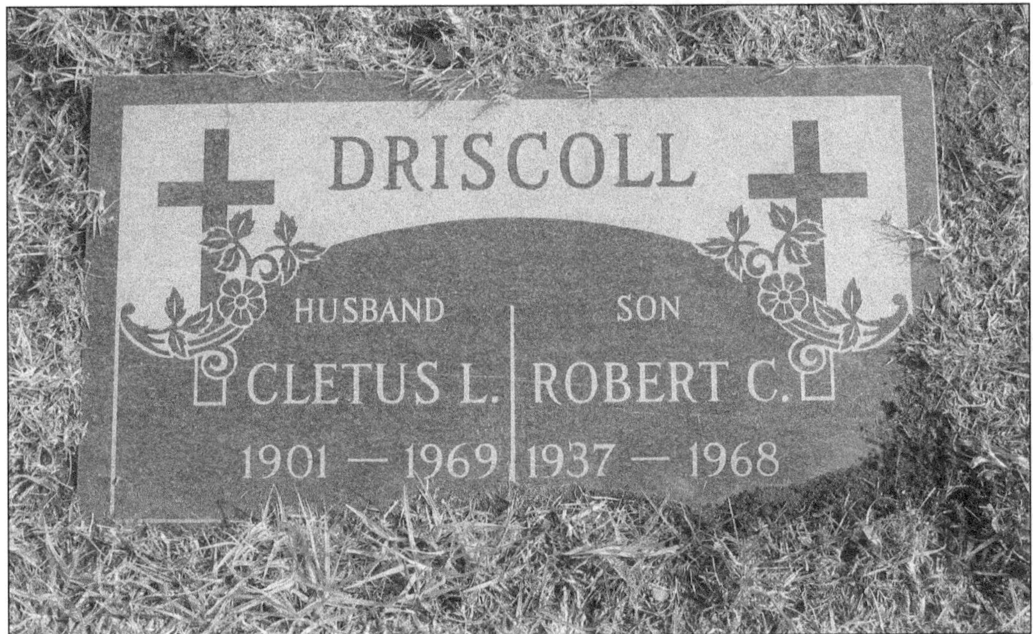

Bobby Driscoll (1937–1968). A highly acclaimed child actor, young Bobby is best known for his roles in *The Window, Song of the South*, and *Treasure Island*. He was also the voice for Disney's *Peter Pan*. His downhill slide of substance abuse ended in a lonely death in a New York tenement. Driscoll was given a pauper's burial in the Potter's Field on Hart Island, New York City. His gravestone at Eternal Hills marks a cenotaph, or empty tomb. (Courtesy SCIC.)

Lloyd Haynes (1934–1987). A television actor with roles in *General Hospital* and *Star Trek*, Haynes is best remembered as the caring history teacher in *Room 222* from the 1960s. Haynes served in both the U.S. Marine Corps and the U.S. Navy. (Courtesy SCIC.)

FALLBROOK CEMETERY FOUNDER. The original Fallbrook Cemetery was located on land homesteaded by Joseph L. Durbin (1815–1887) and set aside as a burial ground at an early date, with the first burial around 1881. Durbin's son had the cemetery officially surveyed following his father's death, adding a Main Street and Laurel Avenue. The three rings on the gravestone indicate his membership in the Independent Order of Odd Fellows. (Courtesy SCIC.)

ODD FELLOWS COMMUNITY CEMETERY TODAY. The Fallbrook Odd Fellows Lodge No. 339 purchased the Fallbrook Cemetery in 1904. The cemetery is non-denominational and holds many of Fallbrook's pioneers; over 1,600 burials have been recorded. A mortuary oversees only the most basic maintenance, as the cemetery is not a perpetual-care facility. (Courtesy SCIC.)

PITTENGER FAMILY, 1895. Among those buried at the Odd Fellows Community Cemetery is Civil War veteran William Pittenger (1840–1904). Pittenger received the Congressional Medal of Honor from Abraham Lincoln in the first such ceremony ever to be conducted, on March 25, 1863. Pittenger served as pastor of the Fallbrook Methodist Church until 1889. (Courtesy Fallbrook Historical Society.)

PITTENGER GRAVESTONE. In 1862, Sgt. William Pittenger was recruited into a military group known as "Andrews Raiders." Posing as Kentuckians, the group infiltrated Confederate territory and hijacked a train in an attempt to disrupt enemy supply lines. For his actions, Pittenger received the Medal of Honor. The operation served as the inspiration for the 1956 Walt Disney film *The Great Locomotive Chase*. (Courtesy SCIC.)

FALLBROOK MASONIC CEMETERY. This is one of five Masonic cemeteries in California and the only one that remains active. It is administered by the Fallbrook Masonic Lodge No. 317, for which it maintains an endowment fund. Although it includes many Masons, the cemetery is non-denominational and is open to the public. (Courtesy SCIC.)

SMELSER GRAVESTONE. Horatio Smelser (1851–1918) and sons founded the Citizens Commercial Bank in Fallbrook. As an active member in the Masonic Lodge, Smelser was charged with locating suitable land for a cemetery in 1916. Fallbrook Masonic Cemetery's first burials took place in 1917. The Masonic Cemetery Association received the deed in 1921. (Courtesy SCIC.)

VALLEY CENTER CEMETERY. The land on which the Valley Center Cemetery is located was part of the original homestead of George Herbst (1822–1883). Rev. J. H. Sherrard purchased the land in the early 1880s and built a house and chapel. The chapel was later converted to a residence that can be seen in this photograph on the right. (Courtesy SCIC.)

DINWIDDIE GRAVESTONE. The Valley Center Cemetery was established in 1883 with the burial of Silence Dinwiddie (1845–1883). George Herbst, the original landowner, died and was interred several months later. Reverend Sherrard donated the cemetery land to the community upon the sale of his house to Dr. James Harrison Clark in 1885. The cemetery is still in use today. However, workers have encountered burials in what should have been empty plots. (Courtesy SCIC.)

ELIZABETH JANE WIMMER, DATE UNKNOWN. "Jennie" Wimmer (1822–1885) is credited as the co-discoverer of California gold at Sutter's Mill on January 24, 1848. A cook and housekeeper on the American River, she was handed a nugget to identify by her husband, Peter, and mill foreman James Marshall. After soaking the "Wimmer Nugget" in a kettle of lye soap, she declared it gold. (Courtesy Valley Center Historical Society.)

WIMMER GRAVESTONE. The Wimmers did not strike it rich during the California Gold Rush. They settled in Valley Center in 1882, near Cole Grade and Cool Valley Roads. Jennie Wimmer died unheralded at age 63 and was buried in the Valley Center Cemetery. Her original grave marker disappeared; for over 50 years, the grave was identified by a brick on which was misspelled, "Weamer." The present plaque was placed in 2003 by E Clampus Vitus. (Courtesy SCIC.)

THE FIRST BETTY CROCKER, C. 1935. Longtime Valley Center resident Agnes White Tizard (1895–1979) was a home economist for the Washburn Crosby Company, manufacturers of Gold Medal Flour. In 1924, she became the radio voice of Betty Crocker on the Betty Crocker Cooking School of the Air, portraying the character on NBC radio for almost 20 years. (Courtesy Valley Center Historical Society.)

TIZARD GRAVESTONE. In 1942, *Fortune* magazine listed Betty Crocker's fame as second only to that of Eleanor Roosevelt. Although the image of Betty Crocker was an artist's compilation of many women, the voice was that of Agnes Tizard. Tizard moved to Valley Center in 1941, where she lived at 29200 Miller Road. (Courtesy SCIC.)

Stokes Ranch Cemetery. The tree-covered knoll to the left of Penn Street in Goose Valley was once the local burial ground for the Ramona area. With the establishment of Nuevo Memory Gardens in 1893, bodies and accompanying gravestones, dating from the 1880s, were transferred to the new cemetery around 1894. Nuevo Memory Gardens can be seen on the hillside in the distance. (Courtesy SCIC.)

Woodson Gravestone. Eula Woodson (1876–1893) was one of the first to be buried in Nuevo Memory Gardens. She was originally interred in the Stokes Ranch Cemetery but exhumed and reburied once the new cemetery was ready. (Courtesy SCIC.)

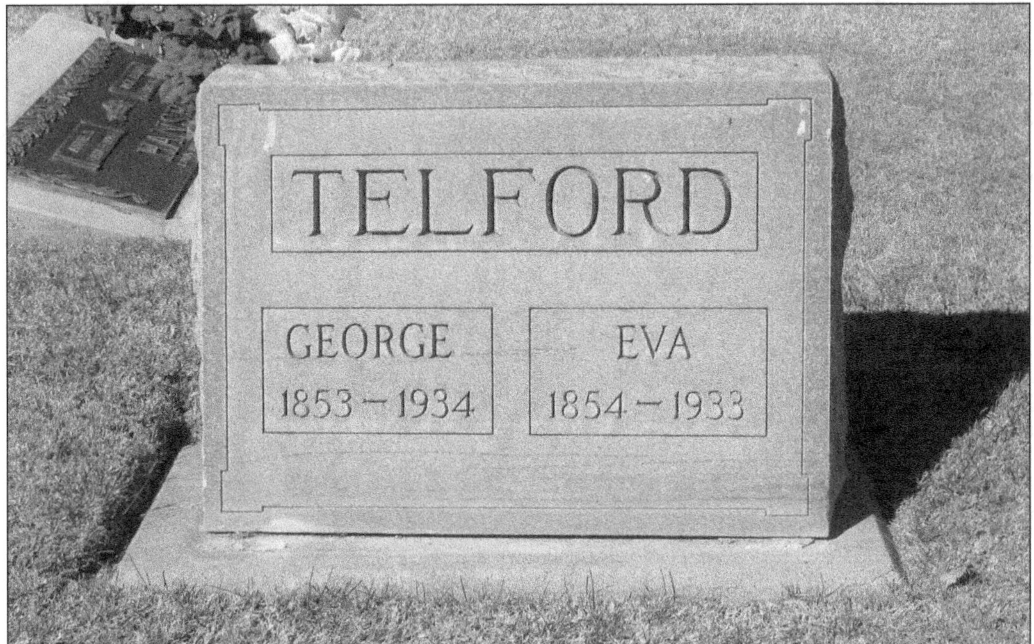

TELFORD GRAVESTONE. George Telford (1853–1934), a Ramona area landowner and farmer, deeded the land for Nuevo Memory Gardens in 1891. Up to that time, locals had no choice but to bury loved ones either on their homesteads or in private burial grounds made available to the public, such as the one on Stokes Ranch. (Courtesy SCIC.)

NUEVO MEMORY GARDENS TODAY. The Nuevo Cemetery Association was incorporated in 1893 to administer Ramona's new cemetery. By the 1930s, it had become increasingly difficult to secure operating funds, but it was not until 1959 that a tax district was formed and improvements were begun, including curbs and pavement, fences, grass, and an irrigation system. One of these improvements, the Ramona War Memorial Gates, was authorized in 1963. (Courtesy SCIC.)

"HAVEN OF REST," C. 1900. Julian's cemetery was established in November 1870 upon the death of John Milton Brockman, aged 14, who was crushed by a tree. It is said that he was to be reburied elsewhere after the winter thaw, but, in fact, his body was never moved. The cemetery plots can be seen on the hillside in the foreground overlooking Main Street. (Courtesy SCIC.)

JULIAN CEMETERY TODAY. This hillside was inaccessible to wagons for over 50 years. Coffins were carried up this walk from Main Street. During the winter of 1896, it reportedly took 16 men to drag the sled bearing Mary Clough's coffin through three feet of snow to her grave. Such manual labor was discontinued in 1924 when vehicle access from A Street was completed. (Courtesy SCIC.)

JOHN M. MCCAIN (1841–1927). McCain is a well-known pioneer name in the East County of San Diego. Patriarch John McCain Sr. and George Hoskings donated the land on which they maintained their family burial plots to the Julian Cemetery Association around 1923. Prior to that, they had allowed the town to use their property as a common burial ground for years. (Courtesy Julian Pioneer Museum.)

MCCAIN GRAVESTONE. The McCains are buried at the highest point in the Julian Cemetery, in the area surrounded by an asphalt road known as the Pioneer Circle. It is the final resting place for approximately 450 individuals, nearly 70 of which are in unmarked graves. (Courtesy SCIC.)

America Newton, c. 1910. An early African American San Diego pioneer, Dyer "America" Newton (1835–1917) is believed to have been a former slave who arrived in Julian from Missouri around 1872 and later homesteaded 80 acres in the area. She earned a living by providing laundry services for the town of Julian through its gold rush days. Newton is buried in the Julian Cemetery, although her grave was unmarked for years. (Courtesy SDHS.)

Black Pioneer Day, 2006. David Lewis of Wynola has spent over seven years researching and mapping the Julian Cemetery. He discovered that a number of graves had been buried or destroyed during the installation of improvements, such as steps and water lines. Lewis is pictured here on the right, making a speech at the 2006 dedication of America Newton's new gravestone. (Courtesy SCIC.)

Conclusion

Best Pal (1988–1998). Golden Eagle Farm in Ramona has a small cemetery dedicated to its horses. Each of the gravestones mark cenotaphs—empty tombs—with one exception; that of their finest racehorse. Best Pal won 18 races and $5,668,245 in purses—a record for a California-breed—before he was retired to Golden Eagle in 1996. His sudden death shocked the racing world. Over 300 fans attended his memorial service. (Courtesy SCIC.)

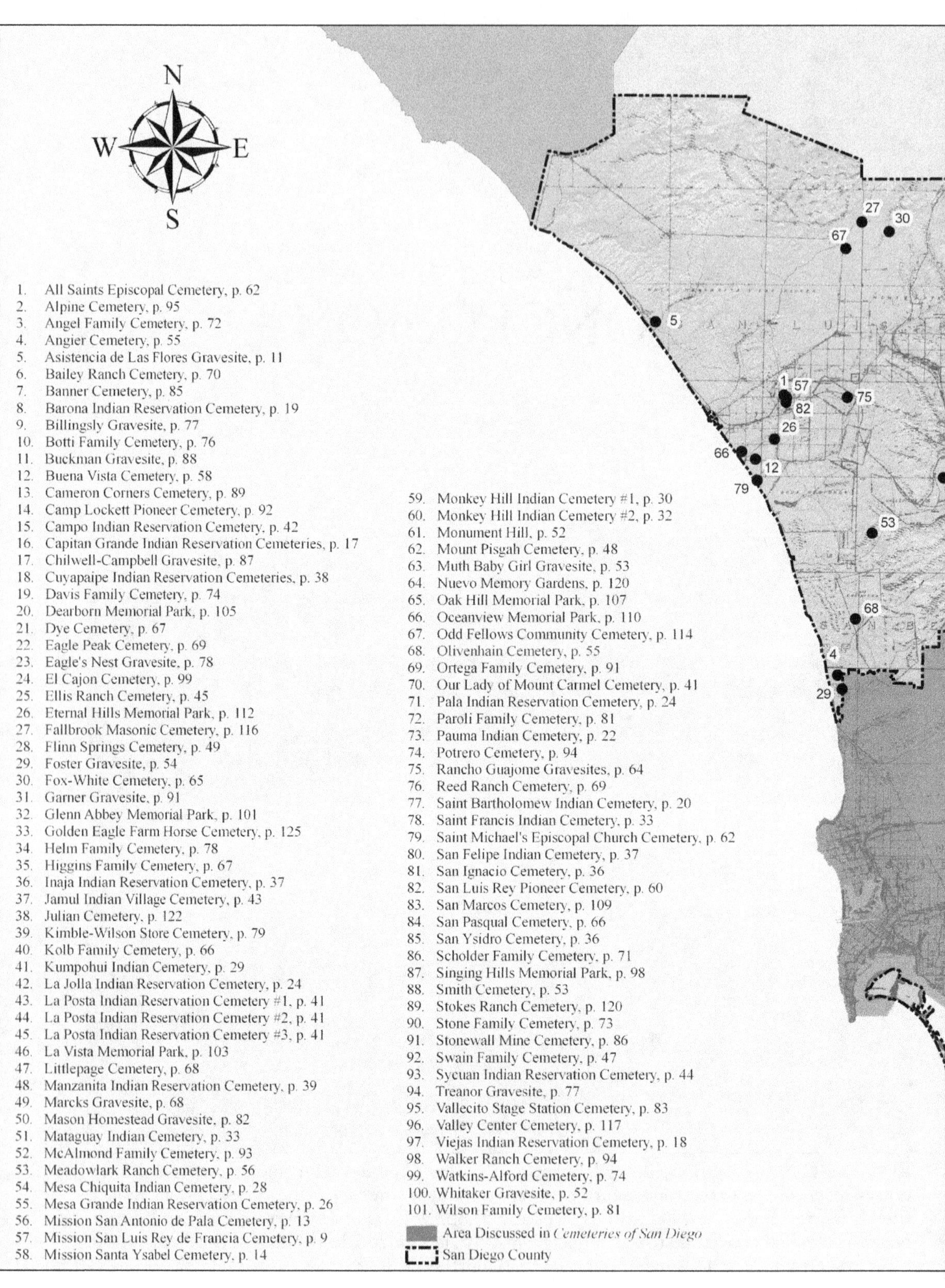
N
W
E
S
1. All Saints Episcopal Cemetery, p. 62
2. Alpine Cemetery, p. 95
3. Angel Family Cemetery, p. 72
4. Angier Cemetery, p. 55
5. Asistencia de Las Flores Gravesite, p. 11
6. Bailey Ranch Cemetery, p. 70
7. Banner Cemetery, p. 85
8. Barona Indian Reservation Cemetery, p. 19
9. Billingsly Gravesite, p. 77
10. Botti Family Cemetery, p. 76
11. Buckman Gravesite, p. 88
12. Buena Vista Cemetery, p. 58
13. Cameron Corners Cemetery, p. 89
14. Camp Lockett Pioneer Cemetery, p. 92
15. Campo Indian Reservation Cemetery, p. 42
16. Capitan Grande Indian Reservation Cemeteries, p. 17
17. Chilwell-Campbell Gravesite, p. 87
18. Cuyapaipe Indian Reservation Cemeteries, p. 38
19. Davis Family Cemetery, p. 74
20. Dearborn Memorial Park, p. 105
21. Dye Cemetery, p. 67
22. Eagle Peak Cemetery, p. 69
23. Eagle's Nest Gravesite, p. 78
24. El Cajon Cemetery, p. 99
25. Ellis Ranch Cemetery, p. 45
26. Eternal Hills Memorial Park, p. 112
27. Fallbrook Masonic Cemetery, p. 116
28. Flinn Springs Cemetery, p. 49
29. Foster Gravesite, p. 54
30. Fox-White Cemetery, p. 65
31. Garner Gravesite, p. 91
32. Glenn Abbey Memorial Park, p. 101
33. Golden Eagle Farm Horse Cemetery, p. 125
34. Helm Family Cemetery, p. 78
35. Higgins Family Cemetery, p. 67
36. Inaja Indian Reservation Cemetery, p. 37
37. Jamul Indian Village Cemetery, p. 43
38. Julian Cemetery, p. 122
39. Kimble-Wilson Store Cemetery, p. 79
40. Kolb Family Cemetery, p. 66
41. Kumpohui Indian Cemetery, p. 29
42. La Jolla Indian Reservation Cemetery, p. 24
43. La Posta Indian Reservation Cemetery #1, p. 41
44. La Posta Indian Reservation Cemetery #2, p. 41
45. La Posta Indian Reservation Cemetery #3, p. 41
46. La Vista Memorial Park, p. 103
47. Littlepage Cemetery, p. 68
48. Manzanita Indian Reservation Cemetery, p. 39
49. Marcks Gravesite, p. 68
50. Mason Homestead Gravesite, p. 82
51. Mataguay Indian Cemetery, p. 33
52. McAlmond Family Cemetery, p. 93
53. Meadowlark Ranch Cemetery, p. 56
54. Mesa Chiquita Indian Cemetery, p. 28
55. Mesa Grande Indian Reservation Cemetery, p. 26
56. Mission San Antonio de Pala Cemetery, p. 13
57. Mission San Luis Rey de Francia Cemetery, p. 9
58. Mission Santa Ysabel Cemetery, p. 14
59. Monkey Hill Indian Cemetery #1, p. 30
60. Monkey Hill Indian Cemetery #2, p. 32
61. Monument Hill, p. 52
62. Mount Pisgah Cemetery, p. 48
63. Muth Baby Girl Gravesite, p. 53
64. Nuevo Memory Gardens, p. 120
65. Oak Hill Memorial Park, p. 107
66. Oceanview Memorial Park, p. 110
67. Odd Fellows Community Cemetery, p. 114
68. Olivenhain Cemetery, p. 55
69. Ortega Family Cemetery, p. 91
70. Our Lady of Mount Carmel Cemetery, p. 41
71. Pala Indian Reservation Cemetery, p. 24
72. Paroli Family Cemetery, p. 81
73. Pauma Indian Cemetery, p. 22
74. Potrero Cemetery, p. 94
75. Rancho Guajome Gravesites, p. 64
76. Reed Ranch Cemetery, p. 69
77. Saint Bartholomew Indian Cemetery, p. 20
78. Saint Francis Indian Cemetery, p. 33
79. Saint Michael's Episcopal Church Cemetery, p. 62
80. San Felipe Indian Cemetery, p. 37
81. San Ignacio Cemetery, p. 36
82. San Luis Rey Pioneer Cemetery, p. 60
83. San Marcos Cemetery, p. 109
84. San Pasqual Cemetery, p. 66
85. San Ysidro Cemetery, p. 36
86. Scholder Family Cemetery, p. 71
87. Singing Hills Memorial Park, p. 98
88. Smith Cemetery, p. 53
89. Stokes Ranch Cemetery, p. 120
90. Stone Family Cemetery, p. 73
91. Stonewall Mine Cemetery, p. 86
92. Swain Family Cemetery, p. 47
93. Sycuan Indian Reservation Cemetery, p. 44
94. Treanor Gravesite, p. 77
95. Vallecito Stage Station Cemetery, p. 83
96. Valley Center Cemetery, p. 117
97. Viejas Indian Reservation Cemetery, p. 18
98. Walker Ranch Cemetery, p. 94
99. Watkins-Alford Cemetery, p. 74
100. Whitaker Gravesite, p. 52
101. Wilson Family Cemetery, p. 81
Area Discussed in *Cemeteries of San Diego*
San Diego County
27
30
67
5
1
57
75
82
26
66
12
79
53
68
4
29

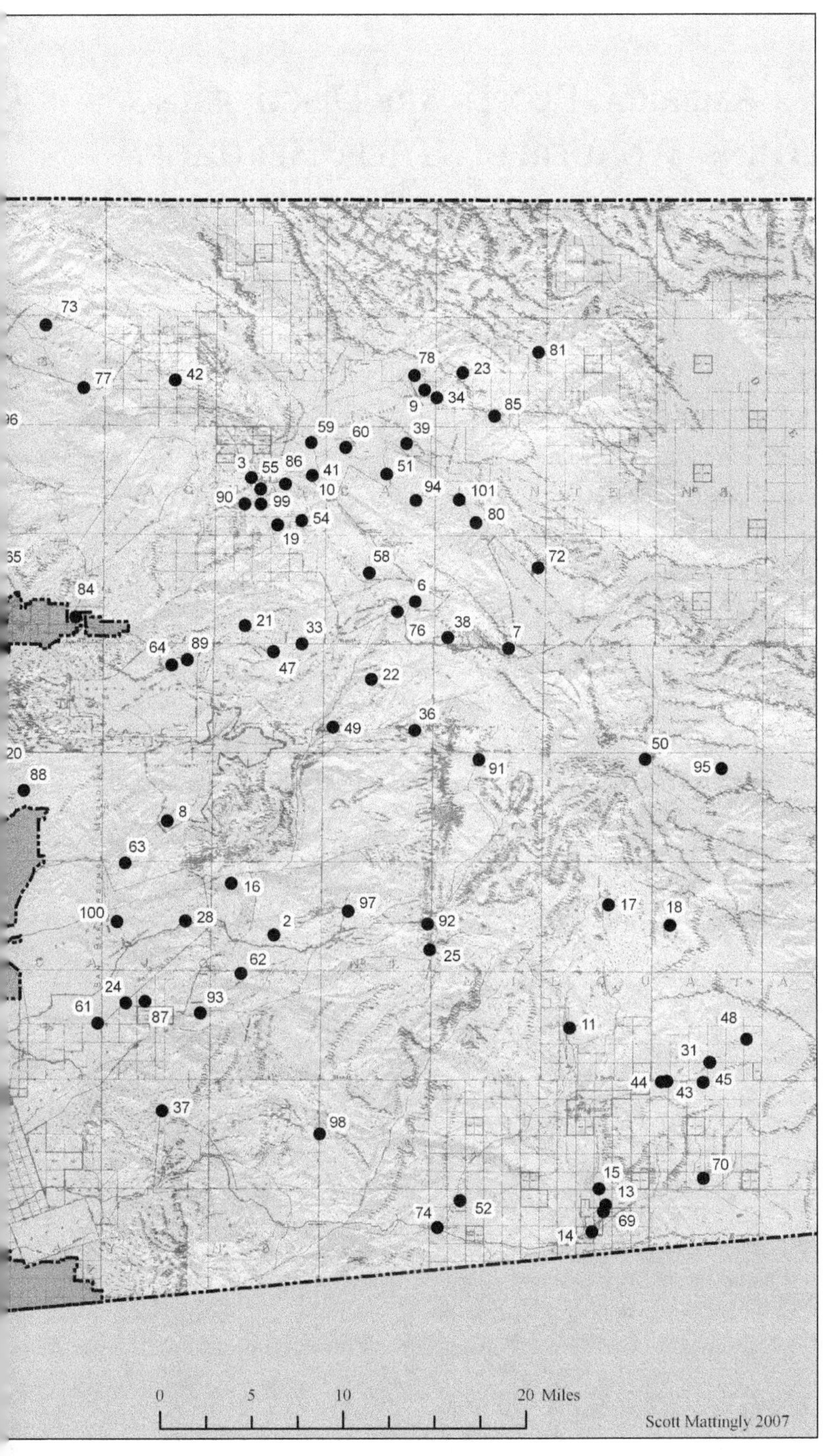

A 19th-Century Map of San Diego. M. C. Wheeler Company's 1872 "Official Map of the Western Portion, San Diego County, California" serves as the backdrop for the exact location of each cemetery discussed in this book. The map covers the territory that is modern San Diego County; what was eastern San Diego County in 1872 is today's Imperial Valley County. (Courtesy Scott Mattingly.)

Across America, People are Discovering Something Wonderful. Their Heritage.

Arcadia Publishing is the leading local history publisher in the United States. With more than 4,000 titles in print and hundreds of new titles released every year, Arcadia has extensive specialized experience chronicling the history of communities and celebrating America's hidden stories, bringing to life the people, places, and events from the past. To discover the history of other communities across the nation, please visit:

www.arcadiapublishing.com

Customized search tools allow you to find regional history books about the town where you grew up, the cities where your friends and family live, the town where your parents met, or even that retirement spot you've been dreaming about.

www.ingramcontent.com/pod-product-compliance
Lightning Source LLC
LaVergne TN
LVHW081532100826
845153LV00004B/259

* 9 7 8 1 5 3 1 6 3 7 4 1 5 *